G-Categories

Recent Titles in This Series

482 **Robert Gordon,** G-categories, 1993

481 **Jorge Ize, Ivar Massabo, and Alfonso Vignoli,** Degree theory for equivariant maps, the general S^1-action, 1992

480 **L. Š. Grinblat,** On sets not belonging to algebras of subsets, 1992

479 **Percy Deift, Luen-Chau Li, and Carlos Tomei,** Loop groups, discrete versions of some classical integrable systems, and rank 2 extensions, 1992

478 **Henry C. Wente,** Constant mean curvature immersions of Enneper type, 1992

477 **George E. Andrews, Bruce C. Berndt, Lisa Jacobsen, and Robert L. Lamphere,** The continued fractions found in the unorganized portions of Ramanujan's notebooks, 1992

476 **Thomas C. Hales,** The subregular germ of orbital integrals, 1992

475 **Kazuaki Taira,** On the existence of Feller semigroups with boundary conditions, 1992

474 **Francisco González-Acuña and Wilbur C. Whitten,** Imbeddings of three-manifold groups, 1992

473 **Ian Anderson and Gerard Thompson,** The inverse problem of the calculus of variations for ordinary differential equations, 1992

472 **Stephen W. Semmes,** A generalization of riemann mappings and geometric structures on a space of domains in $\mathbf{C}^n$, 1992

471 **Michael L. Mihalik and Steven T. Tschantz,** Semistability of amalgamated products and HNN-extensions, 1992

470 **Daniel K. Nakano,** Projective modules over Lie algebras of Cartan type, 1992

469 **Dennis A. Hejhal,** Eigenvalues of the Laplacian for Hecke triangle groups, 1992

468 **Roger Kraft,** Intersections of thick Cantor sets, 1992

467 **Randolph James Schilling,** Neumann systems for the algebraic AKNS problem, 1992

466 **Shari A. Prevost,** Vertex algebras and integral bases for the enveloping algebras of affine Lie algebras, 1992

465 **Steven Zelditch,** Selberg trace formulae and equidistribution theorems for closed geodesics and Laplace eigenfunctions: finite area surfaces, 1992

464 **John Fay,** Kernel functions, analytic torsion, and moduli spaces, 1992

463 **Bruce Reznick,** Sums of even powers of real linear forms, 1992

462 **Toshiyuki Kobayashi,** Singular unitary representations and discrete series for indefinite Stiefel manifolds $U(p,q;\mathbb{F})/U(p-m,q;\mathbb{F})$, 1992

461 **Andrew Kustin and Bernd Ulrich,** A family of complexes associated to an almost alternating map, with application to residual intersections, 1992

460 **Victor Reiner,** Quotients of coxeter complexes and P-partitions, 1992

459 **Jonathan Arazy and Yaakov Friedman,** Contractive projections in C_p, 1992

458 **Charles A. Akemann and Joel Anderson,** Lyapunov theorems for operator algebras, 1991

457 **Norihiko Minami,** Multiplicative homology operations and transfer, 1991

456 **Michał Misiurewicz and Zbigniew Nitecki,** Combinatorial patterns for maps of the interval, 1991

455 **Mark G. Davidson, Thomas J. Enright and Ronald J. Stanke,** Differential operators and highest weight representations, 1991

454 **Donald A. Dawson and Edwin A. Perkins,** Historical processes, 1991

453 **Alfred S. Cavaretta, Wolfgang Dahmen, and Charles A. Micchelli,** Stationary subdivision, 1991

452 **Brian S. Thomson,** Derivates of interval functions, 1991

451 **Rolf Schön,** Effective algebraic topology, 1991

450 **Ernst Dieterich,** Solution of a non-domestic tame classification problem from integral representation theory of finite groups ($\Lambda = RC_3, v(3) = 4$), 1991

(*Continued in the back of this publication*)

MEMOIRS
of the
American Mathematical Society

Number 482

G-Categories

Robert Gordon

January 1993 • Volume 101 • Number 482 (first of 4 numbers) • ISSN 0065-9266

American Mathematical Society
Providence, Rhode Island

1991 *Mathematics Subject Classification.*
Primary 18A35, 18A40, 18B25, 18C15, 18D05, 18E05.

Library of Congress Cataloging-in-Publication Data

Gordon, Robert, 1935–
G-categories/Robert Gordon.
p. cm. – (Memoirs of the American Mathematical Society, ISSN 0065-9266; no. 482)
Includes bibliographical references.
ISBN 0-8218-2543-7
1. Categories (Mathematics) I. Title. II. Series.
QA3.A57 no. 482
[QA169]
510 s–dc20
[511.3]

92-33390
CIP

Memoirs of the American Mathematical Society

This journal is devoted entirely to research in pure and applied mathematics.

Subscription information. The 1993 subscription begins with Number 482 and consists of six mailings, each containing one or more numbers. Subscription prices for 1993 are $336 list, $269 institutional member. A late charge of 10% of the subscription price will be imposed on orders received from nonmembers after January 1 of the subscription year. Subscribers outside the United States and India must pay a postage surcharge of $25; subscribers in India must pay a postage surcharge of $43. Expedited delivery to destinations in North America $30; elsewhere $92. Each number may be ordered separately; *please specify number* when ordering an individual number. For prices and titles of recently released numbers, see the New Publications sections of the *Notices of the American Mathematical Society.*

Back number information. For back issues see the *AMS Catalog of Publications.*

Subscriptions and orders should be addressed to the American Mathematical Society, P. O. Box 1571, Annex Station, Providence, RI 02901-1571. *All orders must be accompanied by payment.* Other correspondence should be addressed to Box 6248, Providence, RI 02940-6248.

Memoirs of the American Mathematical Society is published bimonthly (each volume consisting usually of more than one number) by the American Mathematical Society at 201 Charles Street, Providence, RI 02904-2213. Second-class postage paid at Providence, Rhode Island. Postmaster: Send address changes to Memoirs, American Mathematical Society, P. O. Box 6248, Providence, RI 02940-6248.

Printed in the United States of America.
This volume was printed directly from author-prepared copy.
The paper used in this book is acid-free and falls within the guidelines
established to ensure permanence and durability. ♾

10 9 8 7 6 5 4 3 2 1 98 97 96 95 94 93

TABLE OF CONTENTS

Introduction vii

1. G-Categories: The Stable Subcategory, G-Limits and Stable Limits 1

2. Systems of Isomorphisms and Stably Closed G-Categories 7

3. Partial G-Sets: G-Adjoints and G-Equivalence 16

4. Par(G-Set) and G-Representability 23

5. Transversals 30

6. Transverse Limits and Representations of Transversaled Functors 40

7. Reflections and Stable Reflections 52

8. G-Cotripleability 63

9. The Standard Factorization of Insertion 73

10. Cotripleability of Stable Reflectors 80

11. The Case of $\mathcal{D}^G$ 88

12. Induced Stable Reflections and their Signatures 98

13. The $\mathcal{D}^G$-Targeted Case 112

References 128

ABSTRACT

A G-category is a category together with an action of a group G on it. This Memoir is a study of the 2-category G-Cat of G-categories, G-functors (functors which commute with the action of G), and G-natural transformations (natural transformations which commute with the G-action), with particular emphasis on the relationship between a G-category and its stable subcategory, the largest sub-G-category upon which G-operates trivially. The motivating example is the G-category of graded modules over a G-graded algebra, in which case the stable subcategory is the module category over the underlying ungraded algebra. Indeed, results given here lead elsewhere to a characterization of G-categories G-equivalent to the G-category of graded modules over a G-graded algebra. The existence of a stronger result involving G-transverse equivalence and thus not expressible in G-Cat, among other more intrinsic grounds, renders ineluctable the study herein of a relaxation, Trans G-Cat, of G-Cat in which G-functors are replaced by functors which commute with the action of G only up to coherent natural isomorphism, and G-natural transformations are adjusted accordingly. The Memoir also contains some very general applications of the theory to various additive G-categories and to G-topoi.

Received by the editor September 19, 1988. Key words and phrases: G-category, G-functor, G-natural transformation, stable subcategory, G-limit, stable limit, stably closed, G-adjoint, G-equivalent, partial G-set, G-representable, transversaled functor, transversaled adjoint, transverse-equivalent, transverse limit, stable reflector, G-cotripleable, G-topos.

INTRODUCTION

A G-category, as spelled out in Section 1, is a category upon which a group G acts. The objects and arrows fixed under the action of G form a distinguished subcategory, called the stable subcategory. The basic concern of this memoir is the relationship between a G-category and its stable subcategory.

G-categories occur frequently, if not by name, in the representation theory of finite dimensional algebras: [1], [8], [13], [15] and [16] to cite a few papers. They have been exploited more systematically in algebraic topology, for instance by Taylor [24] and Fiedorowicz, Hauschild and May [6]. The former article speaks of "G-groupoids" while, in the latter, certain topological categories are termed "G-categories".

Regarding the definition given by Fiedorowicz et al., it should be mentioned that G-categories may be defined as category objects in the topos G-Set. We do not take this point of view for the same reason we do not adopt any of the other many nonelementary equivalent notions of a G-category: We wish this memoir, in the light of its possible applications to algebra (see thereto Gordon [9], [11]) and topology, to be readable by any algebraist or topologist with a good basic knowledge of category theory such as provided by Mac Lane [22].

Nevertheless, especially since the sequel [14] to the memoir will be couched entirely in 2-categorical language, it seems appropriate to remark here that viewing G as a one-object category, G-categories are objects of the functor-2-category (G,Cat). This 2-category is herein denoted G-Cat. It is introduced in Section 5 as the sub-2-category of a 2-category Trans-Cat determined by all objects of Trans-Cat, those arrows that are G-functors, and those 2-cells that are G-natural transformations. These arrows and 2-cells amount precisely to functors and natural transformations that commute with the action of G. Thereby, Trans-Cat may equally be described as the relaxation of G-Cat gotten by replacing G-functors by functors that commute with the operation of G up to coherent in G natural isomorphism, and adjusting 2-cells accordingly. Such functors we call transversaled functors. In 2-categorical notation, Trans-Cat is Ps(G,Cat) -- see [3] wherein the corresponding notation is Psd[G,Cat].

Alternately, G-Cat and Trans-Cat may be described as follows, as is used extensively in [14]. Regarding G as a (discrete) monoidal category, G-Cat is the 2-category of strict algebras for the (strict) monad G x — on Cat with strict algebra maps and the usual algebra 2-cells, while Trans-Cat is the 2-category of strict algebras for the same monad, but with pseudo-maps of algebras (called "strong" by Kelly and Street [20, pp. 95 and 96] -- cf. [4]) and the evident adjustment of 2-cells. Indeed, these descriptions make immediately available for G-Cat and Trans-Cat very recent results of Blackwell, Kelly and Power [4]. Thus insertion of G-Cat in Trans-Cat has a left 2-adjoint [20]. Also although Trans-Cat, unlike G-Cat, is not complete, it does have many 2-categorical limits (see [19], [3]): for example cotensors, equifiers, iso-inserters, inserters, and all bicolimits.

Finally, in keeping with the discussion at the outset, G-Cat is Cat(E), the 2-category of internal categories in a topos E for E = (G,Set). Now the 2-categorial aspects of Cat(E) are vital to us; and not many papers overtly take the 2-categorical viewpoint, Street's article [23] being an exception. Thereto, a perspective more geometric than, say, Street's is afforded by the internal logic of a topos [21]. Yet, from either algebraic or geometric standpoint, the simplest possibility we see of studying Trans-Cat in terms of Cat(E) occurs when E at very least has the form (S,Set) for an arbitrary small category S.

Our present elementary approach is made possible chiefly by the introduction, in Section 3, of the G-category of "partial G-sets". This G-category, denoted par(G-Set), plays a role relative to G-categories closely analogous to that played by Set relative to ordinary categories. Technically speaking, inasmuch as par(G-Set) is equivalent to Set (see §4), it is a symmetric monoidal closed category [18]; and any G-category has a built in canonical partial G-set structure on its hom-sets (§3) which in fact makes it the free par(G-Set)-category [18] on its underlying ordinary category. Moreover, because the set-valued hom-functor par(G-Set)(+, —), where + is the unit object of par(G-Set) (effectively the one-point trivial G-set), is obviously faithful and reflects isomorphisms, one can ignore as we do in the text the enriched category theory of par(G-Set)-Cat. For example, it is no more necessary to study nonconical limits [18, Chapter 3] in G-categories than it is to study them in ordinary categories.

The first half of the memoir -- through Section 6 -- deals with the fundamental properties of G-categories. From the notions of G-functor and G-natural transformation already described, one gets a notion of a G-cone to a G-diagram in a G-category, and thus a notion of a G-limit of a G-diagram as a limiting G-cone. On the other hand, the stable limit of a G-diagram Z is a representation, in terms of the aforementioned par(G-Set), of the 1-cell cone(—,Z) in G-Cat. This, and its ramifications, are discussed in Section 4 while, in Section 1, stable limits are defined as limiting cones which are G-cones and shown to be G-limits. Also in Section 1, stably closed G-categories appear as G-categories in which every G-diagram that has a limit as ordinary diagram has a stable limit.

In Section 2, stably closed G-categories are studied by means of systems of isomorphisms; that is, G-indexed families of arrows of the form ${}^{g}X \to X$ (for a fixed object X) satisfying the evident coherence rules. Systems of isomorphisms can be construed as G-diagrams indexed by a certain G-groupoid. The resultant stable limits are readily calcuable (see 2.2) and turn out to exist precisely when the category is stably closed -- see Theorem 6.7 for the more cumbersome case of G-limits. In particular, the property of being stably closed is self-dual. Also, systems of isomorphisms make it possible in practice to recognize, as in Theorem 7.4, when a G-category is stably closed.

In a different vein, when the stable subcategory is reflective, systems of isomorphisms enable one to work out the structure of algebras for the monad generated by the underlying adjunction -- see Section 9. If the category has stable G-indexed coproducts, one finds it to be stably closed if and only if insertion of its stable subcategory is tripleable. Because stable limits are G-limits, the existence of stable G-indexed coproducts implies the stable subcategory is reflective (see 7.1); and then one says it is stably reflective.

We refer the reader to [9] for all manner of examples and counterexamples concerning the stable subcategory, systems of isomorphisms and so on. A few of these are mentioned in the text. We should remark [9] is based on the observation that category objects in Grp are G-categories under conjugation by elements of their object groups G. Here the stable subcategory is readily identifiable and, most interestingly, the G-categories themselves are stably closed precisely when, viewed as crossed modules, every derivation belonging to a certain normal subgroup of the Whitehead group of derivations is principal. For instance, the internal category in Grp corresponding to a G-module M (this seen as trivial crossed module) is stably closed as G-category when and only when the one-dimensional cohomology group $H^1(G,M)$ is trivial.

Naturally G-adjointness and transverse-adjointness; that is, adjointness in the respective 2-categories G-Cat and Trans-Cat, are important concepts concerning G-categories. Having shown, in Section 3, that right G-adjoints are G-continuous, in Section 4 it is shown exactly when par(G-Set)-valued partial G-hom-functors are G-continuous. Not until Section 6 is it demonstrated that right transversaled adjoints preserve transverse limits (i.e., Trans-Cat limits). Also, regarding G-adjoints, a result of apodictic interest to 2-categorists is presented in Section 3; namely, a G-adjoint functor theorem.

Also, in Section 3 are introduced the hereditarily stable closed G-categories we find a focal point of G-category theory: their defining property is simply that they are H-stably closed for every subgroup H of G. For example, not surprisingly from the discussion above, the G-category corresponding to a G-module M is hereditarily stably closed if and only if $H^1(H,M) = 0$ for all subgroups H. Another example of an hereditarily stably closed G-category is par(G-Set) -- see Theorem 4.3. Incidentally, that hereditarily stably closed G-categories are abundant is attested to by Theorems 7.4 and 7.6; in fact, thereby every category (resp. every category with G-indexed coproducts) can be seen as the stable subcategory (resp. stably reflective stable subcategory) of an hereditarily stably closed G-category.

Section 5 is concerned with transversaled adjoints. Although some of our results are valid in Ps(S,Cat) for any small category S via general 2-categorical considerations, others are not valid there. For instance, the generalization to Ps(S,Cat) of the result asserting a transversaled functor having an adjoint has a transversaled adjoint is invalid for most small S. The positive result, Proposition 5.2, is important since typically not even a G-functor which is an equivalence (thus, by 5.2, a transverse-equivalence) need have a left or right G-adjoint; indeed, this is a central issue in the article [11]. Yet, as observed in Section 6, a transversaled functor between hereditarily stably closed G-categories that has an adjoint is naturally isomorphic, in the Trans-Cat sense, to a G-functor. This is because the result, Theorem 3.12, that G-functors with hereditarily stably closed domains that have adjoints have G-adjoints extends to transversaled functors -- cf. Theorem 5.6.

As one might adjudge from the foregoing, lifting G-adjoints can present a thwarting problem unless one happens to be lifting to a G-functor with hereditarily stably closed domain -- a frequently used tactic in Section 13 (see 13.9 and comments thereafter). Of course, by the above cited Proposition 5.2, lifting transversaled adjoints causes no more trouble than lifting ordinary ones. For these, and other, reasons we believe that from the strictly categorical point of view it is better to work in Trans-Cat than in G-Cat, particularly with the knowledge that results valid in the former may be transferred to the latter in the hereditarily stably closed case (as done in [11]). In [14] we have in fact opted to work in Trans-Cat. However, in the second half of this memoir, we elect to remain in G-Cat and work basically with hereditarily stably closed G-categories. This we do both because it is more elementary and straightforward and because it appears to be where the balance of algebraic applications live. The sole exception is a brief excursion in Section 7 concerning the Trans-Cat analogue, the evident forgetful functor from trans$C$ (§5) to C, of the stable subcategory of a G-category C. (This extension of the notion of stable subcategory will become prominent in [14].)

More precisely, the second half of the memoir deals mainly with stable reflections on a G-category, say, C. The corresponding reflectors -- called stable reflectors -- are cotripleable whenever C is an amenable (= idempotents split) Ab-category or, on entirely different grounds, when C is a topos. The former result is a special case of Theorem 10.12 while the latter one is part of Theorem 13.10. In between, in Section 12, a stable reflector on a G-category B and G-cotripleable (= cotripleable in G-Cat) functor $F: A \to B$ are given, and the consequences of this arrangement analyzed in detail. In particular, it is found that stable reflectors on A exist and each is related to the given stable reflector on B by an invariant (of sorts) called its signature -- naturally, in many cases, cotripleability of stable reflectors ascends from B to A (see 12.4 and 12.8).

Subsequently, in the terminal section, the situation just described is specialized to the case of the hereditarily stably closed functor-G-category $B = \mathcal{D}^G$ -- already studied in Section 11 -- where $\mathcal{D}$ is any category with G-indexed coproducts and G is seen as discrete G^{op}-category. This choice of B forces A to be hereditarily stably closed (see 8.5) and, of course, to possess G-indexed coproducts. Here the given stable reflector for B is taken to be the colimiting functor $L: \mathcal{D}^G \to \mathcal{D}$. Notice that if $\mathcal{D}$ is an Ab-category with a non-null object, and G is nontrivial, although L is left adjoint to the diagonal G-functor $\Delta: \mathcal{D} \to \mathcal{D}^G$, L cannot be chosen to be a left G-adjoint of Δ. This follows from the fact that no two of the injections for a nontrivial G-fold copower in $\mathcal{D}$ can be equal.

If $\mathcal{D}$ is additive and amenable and F preserves finite products, the stable subcategory of A, as well as A, are shown to be additive and amenable. If $\mathcal{D}$ is a topos and F is left exact, essentially by a result of F.W. Lawvere and M. Tierney (cf. 13.9) familiar to topos theorists, A is a topos; moreover, the stable subcategory of A is a topos and its inclusion functor logical. The circle is completed at the end of the section by establishing that if C is an hereditarily stably closed G-category with G-indexed products and coproducts, and if C is either additive and amenable or a topos, then C is the domain of a $\mathcal{D}^G$-valued G-cotripleable functor for some category $\mathcal{D}$ with G-indexed products and coproducts.

The classic example of the foregoing is $\mathcal{D}$ = Ab, A = category of externally graded modules over an externally G-graded ring Λ, and F = functor that forgets the Λ-action. Notice that the stable subcategory of A is the module category over the underlying ungraded ring $\underline{\Lambda}$ of Λ. Now, in this setup, the gradable $\underline{\Lambda}$-modules are of profound significance (e.g. [12], [13], [15]); and, in categorical terms, are just the objects in the replete image of a stable reflector on A. Thus, because this stable reflector is cotripleable, finding the gradable modules amounts precisely to determining the underlying objects of the coalgebras for the comonad generated by the adjunction. So guided by the classic example, this forms a recurrent theme in the last several sections of the memoir. When $\mathcal{D}$ is additive and amenable, G-fold copowers in $\mathcal{D}$ embed in G-fold powers, and F preserves finite products, the replete image of a stable reflector for A is characterized in Theorem 13.7 in terms of the signature and the existence of stable sections of components of the counit of reflection giving rise to certain G-indexed families of orthogonal idempotents in $\mathcal{D}$. When $\mathcal{D}$ is a topos and F is left exact, a yet simpler characterization is obtained -- see Theorem 13.8.

Returning to the case when $\mathcal{D}$ = Ab and A = graded Λ-modules, of course, A is hereditarily stably closed. The striking thing is that, roughly speaking, A is characterized among G-categories equivalent to (ungraded) module categories by this property. To make this precise, regarding $\mathcal{D}^G$ as the monoidal category of G-graded abelian groups, note one may form for any $\mathcal{D}^G$-category Λ the evident G-category Λ-Mod = $[\mathcal{D}^G\text{-Cat}](\Lambda,\mathcal{D}^G)$ of "Λ-modules" -- see [10], [11]. (This Λ-Mod reduces, when Λ has a single object, to the ordinary graded Λ-module category A.) In [11] it is shown that if a G-category is abelian, cocomplete and has a small generating set of small projectives, then it is G-equivalent to Λ-Mod for some $\mathcal{D}^G$-category Λ precisely when it is hereditarily stably closed. This is done using results of this memoir and [10], and provides a glimpse of the enriched G-category machinery to be developed in [14].

Briefly, if $\mathcal{D}$ is either a complete topos or a Grothendieck category and carries a symmetric monoidal closed structure (e.g. the cartesian one in the topos case), by results of [14], $\mathcal{D}^G$ is an involution-symmetric monoidal closed G-category (see also [11]) in such a way that the left adjoint L defined above becomes a monoidal adjoint. In particular, given a $\mathcal{D}^G$-category Λ, the $\mathcal{D}$-category $\underline{\Lambda} = L_*\Lambda$, for the L_* of Eilenberg and Kelly [5, p.469], may be formed. Then the category $\underline{\Lambda}$-Mod = $[\mathcal{D}\text{-Cat}](\underline{\Lambda},\mathcal{D})$ may be identified with the stably reflective stable subcategory of the hereditarily stably closed G-category Λ-Mod = $[\mathcal{D}^G\text{-Cat}](\Lambda,\mathcal{D}^G)$. Insertion of this stable subcategory is tripleable. Moreover, the stable reflector is cotripleable; enabling one if one will to seek out the "gradable" $\underline{\Lambda}$-modules as previously outlined.

Plainly, here, applications both algebraic and categorical are fully alive in cases other than when $\mathcal{D}$ consists of possibly adorned modules over a communtative ring R (such as $\mathbb{Z}$-graded or differential-$\mathbb{Z}$-graded ones). Thus $\mathcal{D}$ might be sheaves of R-modules over a topological space, sheaves of sets over a topological space or, indeed, a complete Grothendieck topos.

Our terminology and notation by and large are those of Mac Lane's treatise [22] wherever it is current. Our limits are small indexed. By the replete image of a functor $F: \mathcal{X} \to \mathcal{Y}$ we mean the full subcategory of $\mathcal{Y}$ given by objects isomorphic to the various objects FX. When the replete image is $\mathcal{Y}$, we say F is replete.

So far as I know, both the discovery of the stable subcategory and the realization of its importance are due to E.L. Green and myself. I wish to thank P. Freyd for numerous conversations relating to this paper, especially during its formative period. Finally, I am deeply indebted to A.J. Power for his help and to S. Mac Lane for his critical reading of an earlier version of this work.

1. G-CATEGORIES: THE STABLE SUBCATEGORY, G-LIMITS AND STABLE LIMITS

A *G-category* is a category X equipped with an action of a group G; that is, with each element g of G is associated an endofunctor ${}^{g}-$ on X such that ${}^{g}-\circ{}^{h}- = {}^{gh}-$ and ${}^{1}- = 1_X$. An object or arrow $*$ of X is called stable if ${}^{g}* = *$ for every $g \in G$. The *stable subcategory*, stab X, *of* X is defined to be the G-subcategory consisting of all stable objects and arrows.

When X is an Ab-category, the action of G is assumed to respect addition of arrows.

Given a G^{op}-category we write the operation of G^{op} as superscripts on the right: $-^{g} = {}^{g^{op}}-$. This makes any $G \times G^{op}$-category X a G-category under conjugation. The resultant G-category is denoted $\dot{X}$ and the G-action is connoted by $\cdot$ too; in other words, $g\cdot * = {}^{g}*^{g^{-1}}$. Since the $G \times G^{op}$-stable subcategory of X shall never be considered, we usually write stab X when we really mean stab $\dot{X}$.

Now, if A and B are G-categories, the functor category B^A is a $G \times G^{op}$-category, where the G-action is induced by the action of G on B and the G^{op}-action is induced by the action of G on A. Objects of stab B^A are called *G-functors* and arrows of stab B^A are called *G-natural transformations*. However, by a *stable functor* or *stable natural transformation* we mean, respectively, a stable object or arrow of B^A viewed as G-category via the action of G on B. In Section 5, G-categories, G-functors and G-natural transformations are seen as forming a sub-2-category of a 2-category whose 1-cells and 2-cells are the respective "transversaled functors" and "transverse natural transformations" defined in that section.

When A is a $G \times G^{op}$-category, there is an additional action of G on B^A -- denoted $\square$ -- that will be used occasionally; namely, $\square$ is the action gotten from the operation of G^{op} on A. Notice that stab B^A (meaning stab $\dot{B^A}$) is a G-subcategory of $\overset{\square}{B^A}$. Hereafter, when stab B^A is being explicitly considered as a G-category, it is with the G-action inherited from the $\square$-action of G on $\mathcal{B}^{\mathcal{A}}$.

Next, by way of defining G-limits, for any G-functor F: $\mathcal{A} \to \mathcal{B}$ there is an evident induced functor stab F: stab $\mathcal{A} \to$ stab $\mathcal{B}$; and we frequently abbreviate stab F to $\underline{F}$. In particular, if $\mathcal{C}$ and J are G-categories with J small, inasmuch as the diagonal functor Δ: $\mathcal{C} \to \dot{\mathcal{C}}^{J}$ is a G-functor, the "stable diagonal" $\underline{\Delta}$: stab $\mathcal{C} \to$ stab $\mathcal{C}^{J}$ is defined. Given a G-functor Z: J $\to \mathcal{C}$ (alias: G-diagram in $\mathcal{C}$ over J), we define a *G-cone to Z* to be an object of the comma category $(\underline{\Delta}{\downarrow}Z)$, a *limiting G-cone* to Z being defined as a terminal object. The limiting object itself we denote $\lim_G Z$ and the expression G-cone (–,Z) stands for the contravariant functor

$$[\text{stab } \mathcal{C}^{J}]\,(\underline{\Delta}-,Z)\colon (\text{stab } \mathcal{C})^{op} \to \mathbf{Set}\ .$$

Of course, by Yoneda, a G-limit of Z (= limiting G-cone) just amounts to a representation of G-cone (–,Z). In Section 4 we shall elaborate further on the interplay of G-limits and representability. An extended notion of "transverse limit" is presented in Section 6.

An ordinary limit of a G-diagram which happens to be a G-cone we term a *stable limit*. We say a G-category $\mathcal{C}$ is *stably closed* if every G-diagram in $\mathcal{C}$ which has a limit has a stable limit. We call a G-category *stably complete* if it has all stable limits and *G-complete* if it has all G-limits. Trivially, a complete stably closed G-category is stably complete and, as we shall see below, any stably complete G-category is G-complete.

Stable completeness requires much more than being stably closed: in the next section we show that to be stably closed a G-category need only have those stable limits indexed by a certain G-groupoid built from G. For example, if $\mathcal{D}$ is any category regarded as trivial G-category (G acts trivially) and G is any group regarded as discrete G-category (objects of G are its elements), then the G-category $\dot{\mathcal{D}}^{G}$ is stably closed. Remarkably, provided it has stable G-indexed products, a G-category $\mathcal{C}$ is stably closed if and only if inclusion I: stab $\mathcal{C} \to \mathcal{C}$ is cotripleable -- see Theorem 9.7. In the example $\mathcal{C}=\dot{\mathcal{D}}^{G}$,

I is cotripleable when and only when it has a right adjoint. This is rather obviously equivalent to C having G-indexed G-products. Since another equivalent statement is that $\mathcal{D}$ has G-indexed products, contripleability of I is actually equivalent to the existence in C of stable G-indexed products. These comments stem from Sections 7 and 8.

We should clarify what G-products and G-equalizers are. The former are defined as G-limits of G-diagrams over a discrete G-category. Put differently, let $\{X_j\}_{j\varepsilon J}$, J a G-set, be a family of objects X_j of a G-category subject to ${}^gX_j = X_{g_j}$ for all g. A G-product of the X_j is a stable object S and J-indexed family of arrows $p_j\colon S \to X_j$ with ${}^gp_j = p_{g_j}$ for all g such that if $p_j'\colon S' \to X_j$ is another family, where S' is stable and ${}^gp_j' = p_{g_j}'$ for all g, then $p_j' = p_ju$ for a unique stable u: $S' \to S$. We write $S = G - \Pi X_j$. Note that, in particular, a G-indexed G-product just consists of an object X and an arrow p: $G\text{-}\Pi_{g\varepsilon G}{}^gX \to X$ such that every arrow from a stable object to X has a unique stable factorization through p. Thus (cf. 1.2) any stable G-indexed product has the form p: $S \to X$ with S stable, where the family $\{{}^gp\colon S \to {}^gX\}_{g\varepsilon G}$ is a product.

As for G-equalizers, consider a stable object S and a G-set W of arrows with domain S. Then the G-equalizer of W is a stable arrow e with codomain S with the property that ue = ve whenever u and v are arrows of W with like codomains, and such that any other stable arrow with this property has a unique stable factorization through e. We remark that the G-equalizer of W may be awkward to compute unless all arrows of W have the same (stable) codomain -- otherwise there may be, for instance, a $w\varepsilon W$ such that, for some $g\varepsilon G$, ${}^g(\text{cod } w) = \text{cod } w$ while ${}^gw \neq w$.

Now if e is the ordinary equalizer of W, e is a stable arrow to the extent that e and ${}^ge$ are equal subobjects of S for any g; this implying that e is a stable equalizer. As a rule we do not make this identification.

G-products and G-equalizers suffice to construct all G-limits. For let Z be a G-diagram over J and let u and v be the stable arrows defined by commutativity of the upper and lower squares for all $a \in \text{arr}J$ in the diagram

$$\begin{array}{ccc} Z_k & = & Z_k \\ \uparrow & & \uparrow \\ G\text{-}\prod_j Z_j & \overset{u}{\underset{v}{\rightrightarrows}} & G\text{-}\prod_{a:j\to k} Z_k \\ \downarrow & & \downarrow \\ Z_j & \xrightarrow[Za]{} & Z_k \end{array} .$$

Then the G-limit of Z is the G-equalizer of u and v (which is just the equalizer of u,v in the stable subcategory).

PROPOSITION 1.1. A G-category is G-complete if and only if it has G-products and G-equalizers of parallel pairs of stable arrows. It is stably complete if and only if it has stable products and stable equalizers of stable parallel pairs. □

The second statement in this result is best understood in the light of the following useful, albeit simple, lemma.

LEMMA 1.2. Stable limits are G-limits.

Proof. Let Z be a G-diagram, and let $\Theta: L \to Z$ be both a limiting cone and a G-cone. Suppose $\Theta': L' \to Z$ is a G-cone. Then, since Θ is a limiting cone, $\Theta \cdot u = \Theta'$ for unique $u: L' \to L$. But then, since Θ is a G-cone

$$\Theta' = g\cdot\Theta' = (g\cdot\Theta)\cdot(g\cdot u) = \Theta\cdot {}^g u$$

for every $g \in G$. So u must be stable. Consequently, Θ is a limiting cone. □

A modest generalization of this result is given in Section 4 and the extension to "transversaled limits" given in Section 6 -- see Theorem 6.9.

COROLLARY 1.3. If a G-diagram Z has a stable limit, then every G-limit of Z is a limit of Z (thus a stable limit). □

Naturally, we call a G-functor which preserves G-limits *G-continuous*. In this terminology:

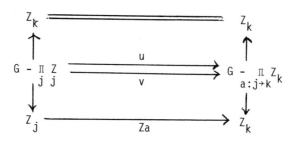

COROLLARY 1.4. Every continuous G-functor from a stably complete G-category

is G-continuous. □

Notice that any continuous G-functor is trivially "stably continuous" in the sense that it preserves all stable limits.

By a *G-trivial limit* we mean a G-limit of a G-diagram over a trivial G-category. In a trivial G-category, every G-trivial limit is a limit and, conversely, every limit can be construed as a G-trivial limit.

We recall that to say a functor $F\colon X \to Y$ creates limits is equivalent to saying a) F creates isomorphisms and b) X has and F preserves limits of diagrams Z in X for which FZ has a limit. In particular, limit creating functors reflect limits too.

PROPOSITION 1.5. Let X be a G-category and denote by I the G-functor given by insertion of stab X in X. Then

i) I creates G-trivial limits;

ii) stab X is as complete as X is G-trivially complete;

iii) I reflects limits.

Proof. The assertion (i) is evident, (ii) is an immediate consequence and (iii) follows from (i) by 1.2. □

Note, in particular, any stable arrow which is an isomorphism in X is an isomorphism in stab X.

Later on, in Lemma 9.2, when stab X is a reflective subcategory of X we prove that insertion stab $X \to X$ is of descent type -- a result which has an overlap with Proposition 1.5. For now, combining the latter with various preceding results, we get

COROLLARY 1.6. The stable subcategory of a stably closed G-category X is as complete as X is, and the inclusion functor preserves the limit of any diagram which has a limit in X. Thus stab X is a complete subcategory of X whenever X is stably complete. □

We end the section with another incisive result concerning creation of limits.

PROPOSITION 1.7. Any G-functor that creates limits creates stable limits.

Proof. Let F be a limit creating G-functor, Z be a G-diagram in dom F and Θ be a stable limit of FZ. Then $F\mu = \Theta$ for a unique cone μ to Z; and μ is a limiting cone. Let $g \varepsilon G$: Since F is a G-functor and Θ is a G-cone

$$\Theta = g \cdot \Theta = g \cdot F\mu = {}^{g}(F\mu)^{g^{-1}} = F({}^{g}\mu^{g^{-1}}) = F(g \cdot \mu).$$

Thus $g \cdot \mu = \mu$, because $g \cdot \mu$ is a cone to Z. Therefore μ is a G-cone. □

This implies that if X is stably complete, any G-category Y admitting a limit creating G-functor $Y \to X$ is stably complete. Actually, more is true; namely, if X is stably closed, so too is Y. We will prove this in the next section -- cf. Lemma 2.11.

2. SYSTEMS OF ISOMORPHISMS AND STABLY CLOSED G-CATEGORIES

Systems of isomorphisms are useful in the study of stable limits. They can be conviently described by starting with the construction of a particular G-groupoid $\mathcal{G}$ from the group G. Objects of $\mathcal{G}$ are elements of G and the action of G on objects is multiplication in G. For each pair of objects g,h, the ordered pair (h,g) is the single arrow from g to h. The action of G on arrows and composition of arrows are forced, to wit: ${}^k(h,g) = (kh,kg)$ and $(k,h) \circ (h,g) = (k,g)$.

Given an object X of a G-category $\mathcal{X}$, a *system of isomorphisms at X* is a G-functor $\mathcal{G} \to \mathcal{X}$ whose value at the identity element of G is X. We mention that for $\mathcal{X}$ a $G \times G^{op}$-category there is a related concept of a "transversal for $\mathcal{X}$" introduced in Section 5. Thereto Section 6 contains more sophisticated versions of certain of the notions and results of the preceding, as well as the present, section. For now we observe that a system of isomorphisms at X is just a G-indexed family $\phi = \{\phi_g\}$ of isomorphisms $\phi_g \in \mathcal{X}({}^gX,X)$ such that the triangle

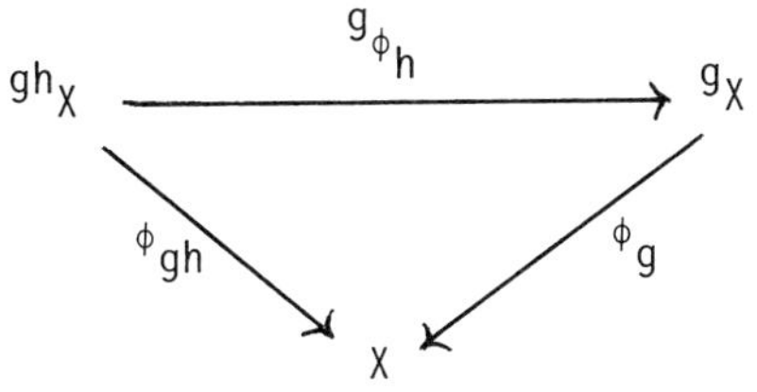

commutes for all $g,h \in G$. We should point out that $\phi_1 = 1_X$ and $\phi_g^{-1} = {}^g\phi_{g^{-1}}$. If X is stable and $\phi_g = 1$ for all g, we say that ϕ is *trivial*. It is worthwhile noting that any family of arrows $\phi_g\colon {}^gX \to X$ such that $\phi_1 = 1$ and the above displayed triangle commutes is a system of isomorphisms. Also, if $\{\phi_g\}$ is a system of isomorphisms, it is readily checked that $\{\phi_g^{-1}\}$ is a cosystem of isomorphisms (i.e., G-diagram over $\mathcal{G}^{op}$) -- we call it the *associated cosystem*.

We refer to stab X^G as the *category of systems of isomorphisms in* X. Concerning this category, the result that follows will be useful later in the memoir. Beforehand we note that if ϕ and Ψ are systems of isomorphisms at, respectively, X and Y, then an arrow ξ: $\phi \to \Psi$ in stab X^G is simply an arrow ξ: $X \to Y$ in X such that the square

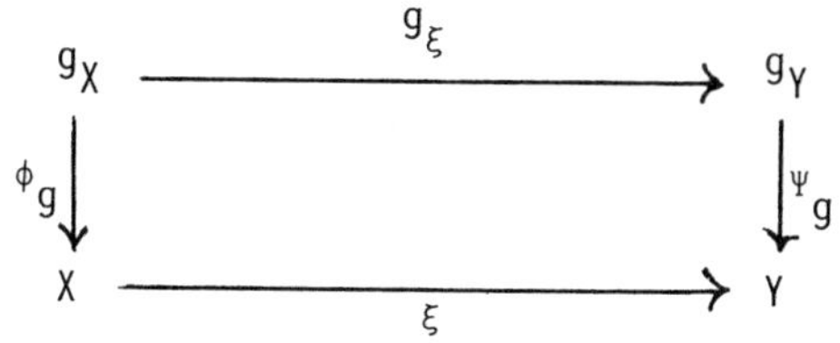

commutes for all g. In particular, we see that stab X^G and stab $X^{G^{op}}$ are isomorphic categories via the assignments

$$\phi_g \longmapsto \phi_g^{-1} \qquad\qquad \xi \longmapsto \xi \quad .$$

THEOREM 2.1. The restriction E_1: stab $X^G \to X$ of evaluation at $1 \in G$ $X^G \to X$ creates limits and colimits.

Proof. Let Z be a diagram in stab X^G over J, and consider the commutative diagram

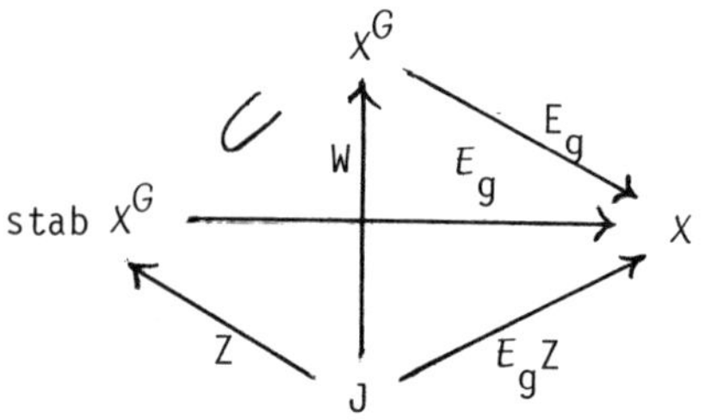

where E_g is evaluation at $g \varepsilon G$.

Let Θ: $X \to E_1Z = E_1W$ be a limiting cone. Then, for every $g \varepsilon G$, $^g\Theta$: $^gX \to {}^g(E_1W)$ is a limiting cone and

$$^g(E_1W) = {}^g(E_1Z) = ({}^gE_1)Z = E_gZ = E_gW.$$

Thus, as is standard, there is a unique cone μ: $F \to W$ ($F \varepsilon X^G$) such that $E_gF = {}^gX$ and $E_g\mu = {}^g\Theta$ for all g; and, moreover, (F,μ) is a limiting cone. Since

$$^h(Fg) = {}^h(E_gF) = {}^{hg}X = E_{hg}F = Fhg$$

F is a G-functor. Since

$$^h(\mu_{jg}) = {}^h(E_g\mu_j) = {}^{hg}\Theta_j = E_{hg}\mu_j = \mu_{jhg}$$

for each $j \in J$, each μ_j is G-natural. Hence (F,μ) may be regarded as a cone to Z and, as such, it is a limiting cone, by 1.5(iii). Also, $E_1F = X$ and $E_1\mu = \Theta$.

Next, suppose $\mu' \colon F' \to Z$ is a cone such that $E_1F' = X$ and $E_1\mu' = \Theta$. We have

$$E_gF' = F'g = {}^g(F'1) = {}^g(E_1F') = {}^gX$$

and, similiarly, $E_g\mu' = {}^g\Theta$. So $F' = F$ and $\mu' = \mu$. Therefore, E_1 creates limits. That E_1 creates colimits follows by duality from the fact that $(\text{stab } X^G)^{op} = \text{stab } X^{opG^{op}}$, since we know stab $X^G \cong$ stab $X^{G^{op}}$ canonically. □

Turning to an analysis of the various types of limits for a system of isomorphisms ϕ at X, we note first that a cone to ϕ is a G-indexed family of arrows, say $\kappa_g \colon Y \to {}^gX$, such that

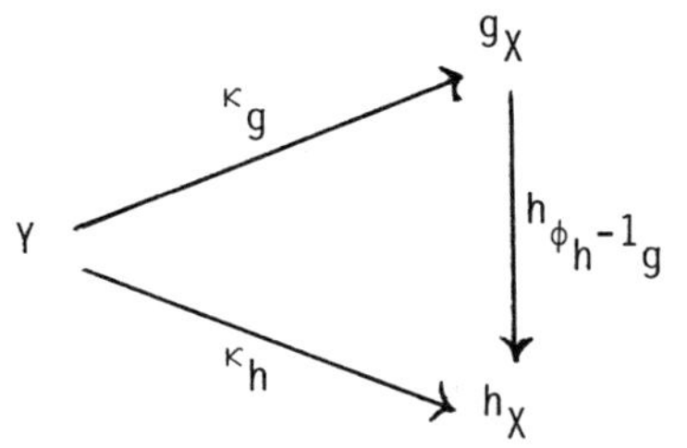

is a commutative diagram. In particular, $\phi_g\kappa_g = \kappa_1$. Conversely, given κ_1 and defining κ_g by $\phi_g\kappa_g = \kappa_1$, since

$${}^h\phi_{h^{-1}g}\kappa_g = {}^h\phi_{h^{-1}g}\phi_g^{-1}\kappa_1 = {}^h\phi_{h^{-1}g}{}^g\phi_{g^{-1}}\kappa_1 = {}^h(\phi_{h^{-1}g}{}^{h^{-1}g}\phi_{g^{-1}})\kappa_1$$

$$= {}^h\phi_{h^{-1}}\kappa_1 = \phi_h^{-1}\kappa_1 = \kappa_h$$

the family $\{\kappa_g\}$ is a cone to ϕ. In other words, any arrow with codomain X is the 1th component of a unique cone to ϕ. It follows that the identity arrow on X is the 1th component of a limiting cone -- the gth component is ϕ_g^{-1}. Now, if $\{\kappa_g\}$ is a G-cone, applying the functor ${}^{h^{-1}}-$ to the above commutative diagram we see that a G-cone to ϕ amounts to an arrow ξ from a stable object S to X rendering commutative the triangle

$$\begin{array}{ccc} & & {}^{g}X \\ & \overset{{}^{g}\xi}{\nearrow} & \downarrow \phi_g \\ S & \underset{\xi}{\searrow} & X \end{array} \qquad (1)$$

Apparently, this ξ is a limit exactly when it is an isomorphism.

To summarize:

LEMMA 2.2. Every system of isomorphisms ϕ at X has a limit, a limiting cone being given by the associated cosystem. A stable limit of ϕ is afforded by a stable object S together with an isomorphism ξ: S $\to$ X subject to commutativity of the triangle (1). □

When ϕ has a stable limit, we say that ϕ is *stabilizing*.

COROLLARY 2.3. A system of isomorphisms is stabilizing if and only if its associated cosystem is costabilizing (has a stable colimit). □

COROLLARY 2.4. An object X of a G-category is isomorphic to a stable object precisely when there is a stabilizing system of isomorphisms at X.

Proof. Let ξ: S $\to$ X be an isomorphism with S stable and put $\phi_g = \xi {}^{g}\xi^{-1}$: ${}^{g}X \to X$. Then

$$\phi_g {}^{g}\phi_h = \xi {}^{g}\xi^{-1} {}^{g}\xi {}^{gh}\xi^{-1} = \xi {}^{gh}\xi^{-1} = \phi_{gh}. \quad □$$

Generally there are systems of isomorphisms at objects isomorphic to no stable object and nonstabilizing systems of isomorphisms at stable objects [9].

COROLLARY 2.5. Let T be a terminal object of a G-category X .

(i) There is a unique system of isomorphisms at T.

(ii) X has a stable terminal object if and only if the system of isomorphisms at T is stabilizing.

Proof. If we denote the arrow ${}^{g}T \to T$ by ϕ_g then, necessarily, $\phi_g {}^{g}\phi_h = \phi_{gh}$ and $\phi_1 = 1$. This proves (i); and then (ii) is an instant consequence of 2.4. □

The first part of the preceding result can be generalized as follows. Suppose Z is any G-diagram admitting a limit, say Θ: X $\to$ Z. Then since, for g$\in$G, g$\cdot\Theta$ is a limit of Z too, for each g there is a unique isomorphism ϕ_g: ${}^{g}X \to X$ making the triangle

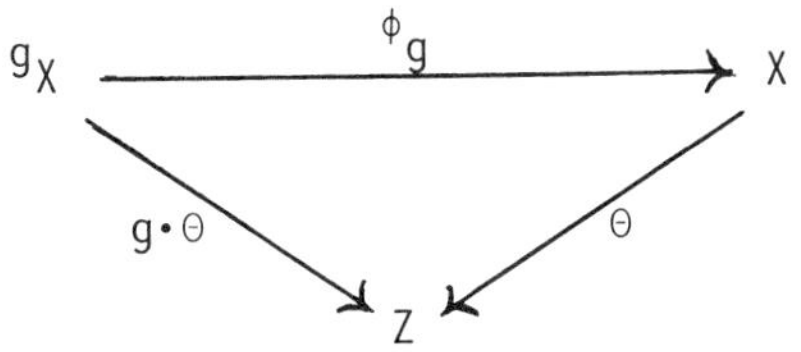

(2)

commute. But then, the ϕ_g are forced by their uniqueness to form a system of isomorphisms. For instance, when Z is a given system of isomorphisms Ψ at X and Θ at g is Ψ_g^{-1}, the induced system of isomorphisms ϕ is just Ψ.

A good deal of the importance of systems of isomorphisms comes from

LEMMA 2.6. Let $\Theta\colon X \to Z$ be a limit of a G-diagram Z and let ϕ be the system of isomorphisms at X defined by commutativity of the diagram in (2). If $\xi\colon S \to X$ is (the 1th component of) a stable limit of ϕ, then $\Theta\cdot\xi\colon S \to Z$ is a stable limit of Z. Conversely, if $\eta\colon S \to Z$ is a stable limit of Z, then the arrow $\xi\colon S \to X$ such that $\Theta\cdot\xi = \eta$ is a stable limit of ϕ. In particular, Z has a stable limit when and only when the system ϕ is stabilizing.

Proof. If ξ is a stable limit of ϕ, then $\Theta\cdot\xi$ is a limit since, by 2.2, ξ is an isomorphism. But

$$g\cdot(\Theta \cdot \xi) = (g \cdot \Theta) \cdot {}^g\xi = \Theta \cdot \phi_g \cdot {}^g\xi = \Theta \cdot \xi$$

by 2.2. Thus $\Theta\cdot\xi$ is a G-cone, hence, a stable limit.

If η is as stated, inasmuch as η is a limit, $\eta = \Theta \cdot \xi$ for an isomorphism $\xi\colon S \to X$. But

$$\Theta\cdot(\phi_g \cdot {}^g\xi) = (\Theta \cdot \phi_g) \cdot {}^g\xi = (g \cdot \Theta) \cdot {}^g\xi = g\cdot(\Theta \cdot \xi) = \Theta \cdot \xi$$

because $\Theta\cdot\xi$ is a G-cone. Therefore $\phi_g \cdot {}^g\xi = \xi$, so that ξ is a stable limit of ϕ, by 2.2. □

This result coupled with Corollary 2.3 yields

THEOREM 2.7. The following are equivalent properties of a G-category X.

(i) X is stably closed.

(ii) Every system of isomorphisms in X is stabilizing.

(iii) X is co-stably closed (G-diagrams in X having colimits have stable colimits). □

Not until Section 6 -- in Theorem 6.7 -- can we determine exactly when every system of isomorphisms has a G-limit.

As an application of the last result, a trivial G-category X is stably closed if and only if for no object X is there a nontrivial morphism of monoids $G \to X(X,X)$. This comment is extended in [14] to arbitrary G-categories in terms of the nonexistence of certain morphisms of G-graded monoids. Here, since we will prove that every category can be viewed as the stable subcategory of a stably closed G-category, our comment makes it plain that the stable subcategory of a stably closed G-category need not itself be stably closed.

The information as to whether a G-category X is stably closed is contained within its category of systems of isomorphisms: evidently, a system of isomorphisms is stabilizing precisely when it is isomorphic to the trivial system of isomorphisms (at the same object). Consequently, by Theorem 2.7, to say that X is stably closed is to say that the full subcategory in stab X^G consisting of the trivial systems of isomorphisms is replete. We denote this full subcategory by triv-stab X^G.

Put otherwise:

COROLLARY 2.8. A necessary and sufficient condition for X to be stably closed is that the stable diagonal $\underline{\Delta}$: stab $X \to$ stab X^G is an equivalence. $\square$

Notice that, if ϕ is a system of isomorphisms at X, ϕ is stable under the $\square$-operation of G if and only if X is stable and $\phi_{gh} = \phi_h \phi_g$ for all g and h; for, since $(h\square\phi)_g = {}^h\phi_{h^{-1}gh}$

$$(h\square\phi)_g = \phi_g \Longleftrightarrow {}^h\phi_{h^{-1}gh} = \phi_g \Longleftrightarrow \phi_h{}^h\phi_{h^{-1}gh} = \phi_h\phi_g \Longleftrightarrow \phi_{gh} = \phi_h\phi_g.$$

In particular, triv-stab $X^G \subset$ stab(stab X^G). In general, however, the inclusion is proper, as we shall see in Section 4.

As we remarked in Section 1, the very definition of stable limits ensures they are preserved by continuous G-functors. The following related result, albeit an easy consequence of Lemma 2.2 and the patent fact that G-functors respect systems of isomorphisms, is surprising.

PROPOSITION 2.9. Any G-functor preserves stable limits (as well as limits) of systems of isomorphisms. □

Next we investigate conditions whereby the systems of isomorphisms at an object X of $\mathcal{X}$ are stabilizing. If $\mathcal{X}$ has G-indexed G-products at X, arr$G$-indexed G-products of the form $G - \prod_{(h,g)}{}^{h}X$, and G-equalizers of parallel pairs of stable arrows, then, as in Proposition 1.1, every system of isomorphisms at X has a G-limit. Now, if it were true that

$$G - \prod_{(h,g)}{}^{h}X \cong G - \prod_{g}(G - \prod_{h}{}^{h}X) \qquad \text{(stable isomorphism)} \qquad (3)$$

the existence of arr$G$-indexed G-products, modulo the existence of the G-indexed G-product on the right, would not be required. However (3) need not be valid unless the G-products concerned are products; and we indicate a proof in that instance: Let $p_{(h,g)}$, q_g and r_h be the projections for the respective stable products $G - \prod_{(h,g)}{}^{h}X$, $G - \prod_{g}(G - \prod_{h}{}^{h}X)$ and $G - \prod_{h}{}^{h}X$, and set $s_{(h,g)} = r_h q_g$. Then since

$$^{k}s_{(h,g)} = {}^{k}r_h{}^{k}q_g = r_{kh}q_{kg} = s_{(kh,kg)} = s_{k_{(h,g)}}$$

the $s_{(h,g)}$ have a stable factorization through the $p_{(h,g)}$. But, in that it is true for ordinary products, the $p_{(h,g)}$ factor through the $s_{(h,g)}$.

In effect what has been shown is:

PROPOSITION 2.10. Every system of isomorphisms at $X \in \mathcal{X}$ is stabilizing provided that $\mathcal{X}$ has stable G-indexed products at X, stable G-fold powers of stable objects and stable equalizers of stable parallel pairs. □

Thus, by Theorem 2.7, if $\mathcal{X}$ has stable G-indexed products and stable equalizers of stable parallel pairs, then $\mathcal{X}$ is stably closed. Yet -- as will be shown in Theorem 9.7 using different methods -- actually it is only necessary for those stable parallel pairs possessing split equalizers in $\mathcal{X}$ to have stable equalizers.

Another consequence of Theorem 2.7, using Proposition 1.7 and the initial part of Lemma 2.2, is that the assertion at the close of the preceding section is valid. However, there is a better result:

LEMMA 2.11. Let X be a stably closed G-category and let $F: Y \to X$ be a G-functor. If either F creates isomorphisms or there is a G-functor $E: X \to Y$ such that $EF \underset{G}{\approx} 1$, then Y is stably closed.

Proof. We use 2.7. For this, let ϕ be a system of isomorphisms at $Y \varepsilon Y$. Then, since F is a G-functor, $F\phi$ is a system of isomorphisms at FY; and then, since X is stably closed, there is a stable limit, say $\sigma: S \to F\phi$.

Suppose F creates isomorphisms. In particular since, by 2.2, $\sigma_1: S \to FY$ is an isomorphism, there is a $T \varepsilon Y$ and an isomorphism $\zeta \in Y(T,Y)$ such that $F\zeta = \sigma_1$. But, for any $g \varepsilon G$, if we apply the G-functor F to the arrow $\zeta^{-1}\phi_g{}^g\zeta: {}^gT \to T$ we get, by 2.2 again, that $\sigma_1^{-1}(F\phi_g)^g\sigma_1 = \sigma_1^{-1}\sigma_1 = 1$. Thus, since F creates isomorphisms,we infer that T is stable and $\zeta^{-1}\phi_g{}^g\zeta = 1$ for all g. Therefore ϕ is stabilizing, by 2.2.

Now suppose that E is as specified and let $\omega: EF \to 1$ be a G-natural isomorphism. But $\omega\phi: EF\phi \to \phi$ is also a G-natural isomorphism and, by 2.9, $E\sigma: ES \to EF\phi$ is a stable limit. Thus $\omega\phi \cdot E\sigma: ES \to \phi$ is a stable limit. So ϕ is stabilizing. □

A direct consequence of this result is that the property of being stably closed is reflected by "G-cotripleable" functors -- see Theorem 8.5(ii). This, in turn, leads to the fact that there are many important stably closed categories -- see [14].

We should explain that the seemingly ad hoc hypothesis asserting the existence of a G-functor E and G-natural isomorphism $EF \to 1$ is really a natural condition. For it is fulfilled whenever F is fully faithful and has a left or right "G-adjoint"; and we define the notion of G-adjointness forthwith: Consider G-functors M and N and an adjunction with unit $\lambda: 1 \to NM$ and counit $\rho: MN \to 1$. If λ and ρ are G-natural, we call this adjunction a *G-adjunction* and say that M is a *left G-adjoint* of N. It should not be surprising were it to suffice for either one of λ,ρ to be G-natural. Indeed, the truth of this will become manifest in the next section where we have more to say about G-adjoints. Here, though, as a useful immediate consequence of the definition, we point out that if M is left G-adjoint to N, then M is left H-adjoint to N for every subgroup H of G. Lastly, we mention that a somewhat different perspective on G-adjointness can be found in Section 5.

3. PARTIAL G-SETS: G-ADJOINTS AND G-EQUIVALENCE

Given a G-set W, the power set of W becomes a G-set if, for any $X \subset W$, one defines

$$^{g}X = \{ \, ^{g}x \mid x \in X \} \qquad (g \in G).$$

By a *partial G-set* we mean a set X together with a G-set W containing X such that $W = \bigcup_{g\in G} {}^{g}X$ (equivalently no proper G-subset of W contains X). A morphism $(X,W) \to (X',W')$ of partial G-sets is a function $X \to X'$; and the category formed of partial G-sets and their morphisms is denoted par(G-Set). An element g of G acts on an object (X,W) by $^{g}(X,W) = (^{g}X,W)$ and on an arrow $f\colon (X,W) \to (X',W')$ by $(^{g}f)^{g}x = {}^{g}(fx)$ for all $x \in X$. Thus par(G-Set) is a G-category. We remark it is not really necessary, in the definition of a partial G-set, to insist that no proper G-subset of W contains X; for, abandoning this stipulation one merely gets a G-category G-equivalent (defined later in the section) to par(G-Set).

The reason for our interest in par(G-Set), other than as a not uninteresting example of a G-category in its own right -- see Section 4 -- is that for any two objects C,D of a G-category C, the hom-set $C(C,D)$ can be viewed as a partial G-set via inclusion in the natural G-set $\bigcup_{g\in G} C(^{g}C,{}^{g}D)$. This entails that $^{g}C(C,D) = C(^{g}C,{}^{g}D)$, so that the partial G-set G-action imparted to arrows of C agrees with the given G-action. It follows, viewing $C^{op} \times C$ as having the evident G-category structure, that the hom-functor $C(-,-)$: $C^{op} \times C \to$ par(G-Set) is a G-functor. By contrast, one should note that the partial one-variable functor $C(C,-)\colon C \to$ par(G-Set), for example, is not a G-functor unless C is stable.

If, more generally, $M\colon P \to Q$ and $N\colon Q \to P$ are G-functors, plainly $Q(M-,-)$ and $P(-,N-)$ are G-functors $P^{op} \times Q \to$ par(G-Set). Suppose, additionally, that M is left adjoint to N with unit λ and counit ρ. Then a short calculation shows that the natural isomorphism Ψ: $\mathcal{Q}(M-,-) \to \mathcal{P}(-,N-)$ defined by $\Psi_{P,Q}q = Nq \circ \lambda_P$, $q \in \mathcal{Q}(MP,Q)$, is G-natural if and only if λ is G-natural. Similarly, Ψ^{-1} is G-natural if and only if ρ is G-natural. But, obviously, Ψ is G-natural if and only if Ψ^{-1} is G-natural. In particular, by the ordinary theory of adjoints, M is left G-adjoint to N in the sense of the definition at the close of the preceding section precisely when there is a G-natural isomorphism of par(G-Set)-valued G-functors, say Ψ: $\mathcal{Q}(M-,-) \to \mathcal{P}(-,N-)$; and then the definition of the action of G on arrows of par(G-Set) leads to the formula

$$^g(\Psi_{P,Q}q) = \Psi^g{}_{{}^gP,{}^gQ}{}^gq, \qquad q \in \mathcal{Q}(MP,Q),\ g \in G.$$

Using this formula one may, by direct computation, easily verify the important fact that right G-adjoints are G-continuous (cf. 6.12). Also we mention a more elaborate version of the above, for "transversaled adjoints", is given in Section 5.

For the next result, we point out that for each $P \in \mathcal{P}$ the comma category $(P\downarrow N)$ is a canonical G(P)-category, where G(P) stands for the stabilizer of P in G. The action of $h \in G(P)$ on an object $(Q,\ p: P \to NQ)$ of $(P\downarrow N)$ is given by $^h(Q,p) = (^hQ, {}^hp)$, while G(P) operates on arrows of $(P\downarrow N)$ in the same way it operates on arrows of $\mathcal{Q}$.

LEMMA 3.1. The G-functor N: $\mathcal{Q} \to \mathcal{P}$ has a left G-adjoint if and only if the G(P)-comma categories $(P\downarrow N)$ have stable initial objects whenever $P \in \mathcal{P}$.

To prove this, making use of the notation above, define M and λ in the evident manner on a representative (with respect to G-action) class of objects of $\mathcal{P}$, and extend to all of obj$\mathcal{P}$ using stability of initial objects of the $(P\downarrow N)$. There is a more general result, valid for "transversaled functors" and requiring a more careful proof, presented in detail later -- see Lemma 5.5.

Of course, Lemma 3.1 implies that if N has a left adjoint and unit of adjunction that are G(P)-stable at every $P \in \mathcal{P}$, then N has a left G-adjoint -- cf. Theorem 3.10. Also, using the next result, it is a simple matter to produce G-functors that have left adjoints but do not have left G-adjoints.

COROLLARY 3.2. If G operates without fixed points on obj$\mathcal{Q}$, then a G-functor $\mathcal{Q} \to \mathcal{P}$ which has a left adjoint has a left G-adjoint exactly when G operates without fixed points on obj$\mathcal{P}$ (equivalently, a G-functor $\mathcal{P} \to \mathcal{Q}$ exists). □

Lemma 3.1 has other immediate applications; significantly, to the notion of "G-equivalence". We call a G-functor N: $\mathcal{Q} \to \mathcal{P}$ a *G-equivalence* if there are G-natural isomorphisms NM ≅ 1 and MN ≅ 1 for some G-functor M: $\mathcal{P} \to \mathcal{Q}$. In general we define the *stably replete image of* N to be the full subcategory of $\mathcal{P}$ with objects those stable objects stably isomorphic to objects of the form NQ for some stable Q. If the stably replete image of N consists of all stable objects of $\mathcal{P}$, we say that N is *stably replete*. When N is H-stably replete for every subgroup H of G, we say that it is *hereditarily stably replete.*

Concerning the interrelatedness of the above terms, we record:

COROLLARY 3.3. The G-functor N: $\mathcal{Q} \to \mathcal{P}$ has the following equivalent properties.

(i) N is a G-equivalence.

(ii) N induces an equivalence $\mathrm{stab}_H\mathcal{Q} \to \mathrm{stab}_H\mathcal{P}$ for all subgroups H of G.

(iii) N is full, faithful, and hereditarily stably replete.

(iv) N is part of a G-adjoint equivalence; that is, N has a left (or right) G-adjoint for which the unit and counit of adjunction are G-natural isomorphisms. □

For instance, when $\mathcal{Q}$ is a trivial G-category, N is a G-equivalence if and only if it induces an equivalence $\mathcal{Q} \to \mathrm{stab}\,\mathcal{P}$ and, for every subgroup H, $\mathrm{stab}_H\mathcal{P}$ is an equivalent subcategory of $\mathcal{P}$.

There will be occasion to be concerned with G-equivalences N that are injective on objects -- we use the term *injective G-equivalence*. Such N are precisely the identity-arrow-reflecting G-equivalences, as will be of use in Section 8. Actually, if F is any fully faithful functor with an adjoint, then F is injective on objects if and only if it reflects identity arrows: for supposing FX = FY and η to be unit and ε counit of an adjunction with F as, say, right adjoint, $F\varepsilon_X^{-1} = \eta_{FX} = \eta_{FY} = F\varepsilon_Y^{-1}$; so $\varepsilon_Y\varepsilon_X^{-1}$ is an arrow $X \to Y$ with $F\varepsilon_Y\varepsilon_X^{-1} = 1$. Another characterization of injective G-equivalences is that they are exactly the G-functors which are equivalences and possess a left G-inverse (i.e., a left inverse G-functor). This is a consequence of the more general

COROLLARY 3.4. The following are equivalent properties of N.

(i) N is injective on objects, fully faithful, and has a left G-adjoint.

(ii) N has a left G-adjoint-left G-inverse.

(iii) N is fully faithful with left adjoint a left G-inverse of N.

Proof. (i)$\Rightarrow$(ii). Let M be left G-adjoint to N with G-natural unit λ. For each $P \varepsilon \mathcal{P}$, define $M'P \;\varepsilon\; \mathcal{Q}$ and $\lambda'_P \;\varepsilon\; \mathcal{P}(P, NM'P)$ by

$$\begin{aligned} M'P &= Q \quad \text{if } P = NQ \text{ for some (necessarily unique) } Q \varepsilon \mathcal{Q} \\ &= MP \quad \text{if } P \notin \text{im}N \\ \lambda'_P &= 1_P \quad \text{if } P \;\varepsilon\; \text{im}N \\ &= \lambda_P \quad \text{if } P \notin \text{im}N \quad . \end{aligned}$$

Then, using full-faithfulness of N, one gets that $(M'P, \lambda'_P)$ is an initial object of $(P{\downarrow}N)$ which, furthermore, is clearly $G(P)$-stable. It follows, as in 3.1, that there is a unique way of defining M' on arrows so that M' becomes a left G-adjoint of N with G-natural unit of adjunction λ'. Let ρ' be the corresponding counit. Since by construction $\lambda'N = 1$, by the relevant triangle identity, $N\rho' = 1$. Thus $\rho' = 1$.

(ii)$\Rightarrow$(iii). Clear.

(iii)$\Rightarrow$(i). Let M be left adjoint-left G-inverse to N with counit ρ. But then, since $N\rho$ is a natural automorphism of N and the triangle

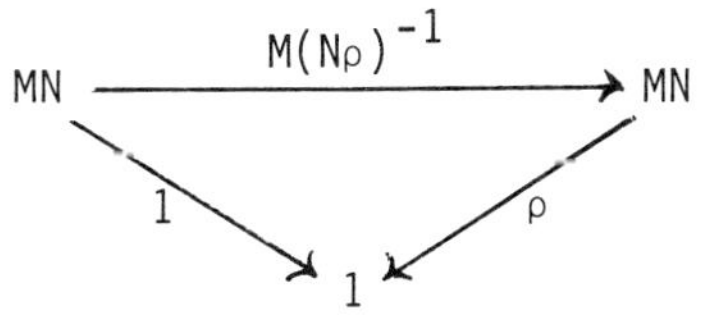

commutes, it follows, inasmuch as $1\colon MN \to 1$ is G-natural, that M is left G-adjoint to N. $\square$

COROLLARY 3.5. Suppose that N is a G-functor and that M is a fully faithful left G-inverse of N. Then N is a G-equivalence and M is left G-adjoint to N.

Proof. The hypothesis implies that M is an equivalence and hence left adjoint to N. □

Next we state our version of the adjoint functor theorem for G-categories.

THEOREM 3.6. Let N: $\mathcal{Q} \to \mathcal{P}$ be a G-functor and suppose that $\mathcal{Q}$ has small hom-sets and is G(P)-complete for all objects P of $\mathcal{P}$. Then N has a left G-adjoint if and only if, for all $P \in \mathcal{P}$, N is G(P)-continuous and the G(P)-category $(P \downarrow N)$ has a weakly initial small set of stable objects.

The proof of this G-adjoint functor theorem is an easy consequence of Lemma 3.1 and the following two results. The proof of the first of these is inspired by Mac Lane's proof of [22, Theorem 1, p.116].

THEOREM 3.7. A G-complete G-category with small hom-sets has a stable initial object if and only if it has a weakly initial small set of stable objects.

Proof. The necessity of the solution set condition is trivial. For the sufficiency since, by 1.5(ii), the stable subcategory of a G-complete G-category has products, we may suppose, say, X is a G-complete G-category with a weakly initial stable object S.

Let e: S' $\to$ S be the G-equalizer of the G-set $X(S,S)$ and let a and b be parallel arrows in X with domain the weakly initial stable object S'. It suffices to show that a = b, so we take the G-equalizer c: E $\to$ S' of the G-set $\{{}^g a, {}^h b \mid g,h \in G\}$. This c is stable and satisfies ac = bc.

Now, from weak initiality of S, there is some arrow ξ: S $\to$ E. Let T be the (stable) domain of the G-equalizer v of the orbit of ξ. In particular there is an arrow w: S $\to$ T; and then, since vw is an endomorphism of S, vwe = e. Thus, for every $g \in G$

$$^g(\xi e) = {}^g\xi\, {}^g e = {}^g\xi e = {}^g\xi vwe = \xi vwe = \xi e$$

that is, ξe is stable.

Next, since $ec\xi \in X(S,S)$, $ec\xi e = e$. But $c\xi e$ is stable and e is monic in stab X. Therefore $c\xi e = 1$ and, consequently, $a = b$. □

Although the proof of the next result, for ordinary categories, is known, we are aware of no suitable reference. Thus we give, as well as the statement, the proof:

LEMMA 3.8. If $N \in \operatorname{stab} \mathcal{P}^{\mathcal{Q}}$, $P \in \mathcal{P}$ and $H < G(P)$, then any H-limit preserved by N is created by the canonical projection $(P{\downarrow}N) \to \mathcal{Q}$.

Proof. Certainly the canonical projection, say F, is an H-functor. Let Z be an H-diagram in $(P{\downarrow}N)$ over J, let $\Theta\colon Q \to FZ$ be an H-limit of FZ such that $N\Theta$ is an H-limit of NFZ, and set $Z_j = (FZ_j, p_j)$, $j \in J$. Then the p_j are the components of a cone $\eta\colon P \to NFZ$. Moreover, inasmuch as Z is an H-functor, η is an H-cone. Thus $\eta = N\Theta \cdot p$ for a unique $p \in \mathcal{P}(P, NQ)$.

Because Θ is an H-cone, F is a faithful H-functor, and (Q,p) is H-stable, $\tau\colon (Q,p) \to Z$ defined by $\tau_j = \Theta_j$ is an H-cone. Obviously, $F\tau = \Theta$. Let $\tau'\colon (Q',p') \to Z$ be another H-cone to Z. Then

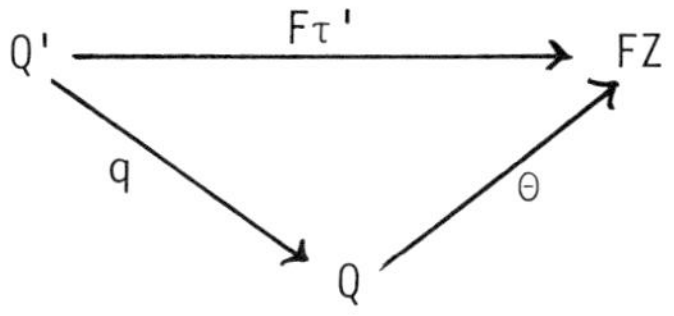

commutes for unique q in Q, and we get the commutative diagram

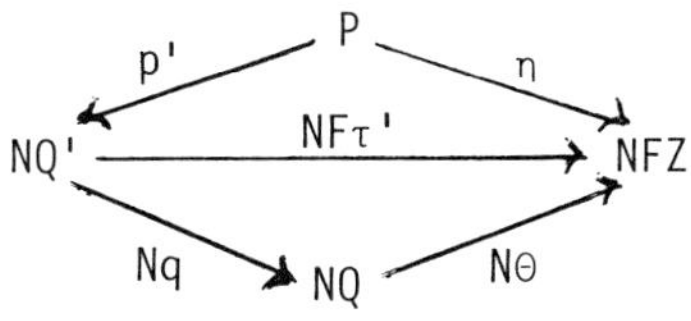

From this we see that $(Nq)p' = p$, so that q is an arrow $(Q',p') \to (Q,p)$ in $(P{\downarrow}N)$. By faithfulness of F, q is in fact the unique such arrow satisfying $\tau \cdot q = \tau'$. Thus τ is a limiting H-cone.

Now, above, when $F\tau' = \Theta$ one gets $q = 1_Q$, so $\tau' = \tau$ in that case. Thus F creates an H-limit for Z. □

We find Theorem 3.6 more difficult to apply than the ordinary adjoint functor theorem. Therefore we seek criteria, more general than the condition of Corollary 3.2, whereby a G-functor, upon having a left adjoint, has a left G-adjoint. Our criteria are based on the following result, which is a direct consequence of the preceding result, Lemma 1.2 and Proposition 2.9.

COROLLARY 3.9. A cosystem of isomorphisms ϕ in $(P{\downarrow}N)$ is costabilizing precisely when the G(P)-cosystem of isomorphisms in $\mathcal{Q}$ gotten by projecting ϕ in $\mathcal{Q}$ is G(P)-costabilizing. □

From this, Corollary 2.5 and Lemma 3.1, we obtain

THEOREM 3.10. Let $M\colon \mathcal{P} \to \mathcal{Q}$ be left adjoint to the G-functor N with unit of adjunction λ. Then N has a left G-adjoint when and only when, for all $P \in \mathcal{P}$, the G(P)-cosystem of isomorphisms $\phi_h\colon MP \to {}^hMP$ in $\mathcal{Q}$ satisfying

$$(N\phi_h)\lambda_P = {}^h\lambda_P \qquad (h \in G(P))$$

is G(P)-costabilizing. □

Thus, insofar as stably closed and co-stably closed are equivalent terms (see 2.7):

COROLLARY 3.11. If N has a left adjoint and $\mathcal{Q}$ is G(P)-stably closed for all P, then N has a left G-adjoint. □

For example, any G-functor from a stably closed G-category to a trivial G-category has a left G-adjoint whenever it has a left adjoint.

We say a G-category is *hereditarily stably closed* if it is a stably closed canonical H-category for all subgroups H of G. Examples in [9] show that a G-category may be a stably closed H-category for every normal subgroup H of G and yet not be hereditarily stably closed.

THEOREM 3.12. A G-functor which has a left adjoint has a left G-adjoint provided its domain is hereditarily stably closed. □

This result will be extended to "transversaled adjoints" in the section after next -- see Theorem 5.7.

4. PAR(G-SET) AND G-REPRESENTABILITY

Just as the notion of representability for ordinary functors revolves around the category Set and its properties, our notion of representability for G-functors revolves around the G-category par(G-Set) and its properties. Thus we take up the latter notion only after first making a study of the pertinent aspects of par(G-Set).

Plainly, stab par(G-Set) = G-Set. More generally, given a subgroup H of G, stab_Hpar(G-Set) is essentially H-Set. The truth of this will evolve from the following argument. Let V be an H-Set. Then, for any G-set W and H-equivariant map f: $V \to W$, $A = \bigcup_{g\in G} {}^g(fV)$ is a G-subset of W containing fV as H-set. Since the cardinality of such A is bounded it follows, by Freyd's adjoint functor theorem, that the forgetful functor G-Set $\to$ H-Set has a left adjoint, say S. Put otherwise, the free G-set SV on V exists; and, by the usual arguments, $V \subset SV$ and $SV = \bigcup_{g\in G} {}^gV$. Notice that the assignments

$$(Y,V) \longmapsto (Y,SV) \qquad\qquad a \longmapsto a$$

define an H-functor par(H-Set) $\to$ par(G-Set). By Corollary 3.5 one has:

LEMMA 4.1. For any subgroup H of G, the H-functor par(H-Set) $\to$ par(G-Set) gotten from the left adjoint of the forgetful functor G-Set $\to$ H-Set as above is an H-equivalence with left H-adjoint-left H-inverse the underlying H-functor U: par(G-Set) $\to$ par(H-Set) sending objects (X,W) to objects $(X, \bigcup_{h\in H} {}^hX)$ and each arrow to itself. □

We regard par(H-Set) as an H-equivalent H-subcategory of par(G-Set) via the injective H-equivalence of this lemma. This makes H-Set = stab_Hpar(H-Set) an equivalent subcategory of stab_Hpar(G-Set). In particular since, taking H = 1, we find Set to be an equivalent subcategory of par(G-Set), par(G-Set) is complete and cocomplete.

Next we argue that par(G-Set) has all H-stable limits and colimits. Along the way we demonstrate that the stable subcategory of stab(par(G-Set))^G (under □) is typically larger than triv-stab(par(G-Set))^G. For this, consider a partial G-set (X,W). If ϕ is a system of isomorphisms at (X,W), G operates on X by $g \circ x = \phi_g {}^g x$. Conversely, if X is a G-set under $\circ$, ϕ_g: ${}^g(X,W) \to (X,W)$ defined by $\phi_g {}^g x = g \circ x$ yields a system of isomorphisms. Thus any G-action $\circ$ on X is induced by a unique system of isomorphisms ϕ at (X,W). Obviously ϕ is trivial if and only if X = W and $\circ$ is the given action of G on X (i.e., on W). On the other hand, we claim ϕ is stable if and only if X = W and the formula

$$g \circ {}^h x = {}^h(h^{-1}gh \circ x)$$

is valid for all g,h and x. But, for all x

$$\begin{aligned} & g \circ {}^h x = {}^h(h^{-1}gh \circ x) \\ \Leftrightarrow\ & \phi_g {}^{gh}x = {}^h(\phi_{h^{-1}gh} {}^{h^{-1}gh}x) = {}^h\phi_{h^{-1}gh} {}^{gh}x \\ \Leftrightarrow\ & (\phi_h \phi_g)^{gh}x = (\phi_h {}^h\phi_{h^{-1}gh})^{gh}x = \phi_{gh} {}^{gh}x \\ \Leftrightarrow\ & \phi_h \phi_g = \phi_{gh} \end{aligned}$$

and the validity of the last equation for all g,h is, as noted in Section 2, equivalent to ϕ being stable.

The assignment of a G-set to each system of isomorphisms yields a canonical functor stab(par(G-Set))^G $\to$ G-Set which is fully faithful. For if ϕ, ϕ' are systems of isomorphisms at (X,W), (X',W') respectively and f: X $\to$ X' is a function, evidently, f is a morphism of systems of isomorphisms $\phi \to \phi'$ exactly when f is equivariant with respect to the corresponding induced G-actions. We have :

PROPOSITION 4.2. The stable diagonal $\underline{\Delta}$: G-Set $\to$ stab(par(G-Set))^G is an equivalence with left adjoint-left inverse the canonical functor sending systems of isomorphisms to the G-sets they induce. □

We say that a G-category is *hereditarily stably complete* if it is a stably complete H-category for every subgroup H of G. The preceding two results, abetted by Corollary 2.8, establish

THEOREM 4.3. Par(G-Set) is hereditarily stably complete and cocomplete. In particular, it is hereditarily stably closed. □

Granted Lemma 4.1, a quicker proof of Thereom 4.3 can be given: if (X_j, W_j) is a family of objects of par(G-Set) indexed by a G-set J and satisfying $({}^gX_j, W_j) = (X_{g_j}, W_{g_j})$ for all $g \in G$, then the cartesian product of the sets X_j becomes a stable product of the (X_j, W_j) under the rule

$$({}^g x)_j = {}^g x_{g_j^{-1}}, \qquad \text{all } x \in \Pi X_j.$$

Thus Theorem 4.3 is a consequence of Proposition 1.1, because par(G-Set) manifestly has stable equalizers of stable parallel pairs. In the same vein we wish to mention that the stable G-indexed coproduct at $(X,W) \in$ par(G-Set) is the free G-set on the set X. This we write $G \cdot X$ since it coincides with the G-fold copower of X (with evident G-action).

Next, consider a G-functor F: $C \to$ par(G-Set). A *G-representation of F* is defined to be a stable object C of C together with a G-natural transformation η: $C(C,-) \to F$ such that η_D is an isomorphism for every stable object D of C (cf. 4.5). A *stable representation of F* consists of a stable object C and a G-natural isomorphism $C(C,-) \to F$. Stable representations are automatically G-representations, but not necessarily conversely; and this is elaborated on below. When F is only an H-functor for some subgroup H of G, *H-representations* and *H-stable representations* are defined to be, respectively, H-representations and stable representations of the composite H-functor

$$C \xrightarrow{F} \text{par(G-Set)} \xrightarrow{U} \text{par(H-Set)}$$

where U is the underlying H-equivalence of Lemma 4.1.

Now, assuming F to be a G-functor, F induces $\underline{F}$ = stab F: stab $C \to$ G-Set. Since, for any arrow γ of C, $F\gamma$ maps fixed points to fixed points, a subfunctor ${}^G F$ of $\underline{F}$ may be defined by taking $({}^G F)C$ to be the set of fixed points of FC. This subfunctor is regarded as Set-valued so that ${}^G -$ becomes a functor $\text{stab(par(G-Set))}^C \to \text{Set}^{\text{stab } C}$ in the evident way. Also, if F is an H-functor, one defines ${}^H F$ to be ${}^H(UF) \in \text{Set}^{\text{stab}_H C}$. For instance, if $F = C(C,-)$ for $C \in \text{stab}_H C$, ${}^H F = [\text{stab}_H C](C,-)$.

LEMMA 4.4 (Yoneda for par(G-Set)-valued G-functors). The Yoneda map

$$[\mathrm{stab}(\mathrm{par}(G\text{-Set}))^{\mathcal{C}}](\mathcal{C}(C,-),F) \longrightarrow ({}^{G}F)C, \qquad \eta \longmapsto \eta_C 1_C$$

is a bijection natural in stable C and F. In particular, there is a canonical one-to-one correspondence between stable representations of F and stable initial objects of the G-category $(\dagger\downarrow F)$, where $\dagger$ stands for the one-point trivial G-set.

Proof. By ordinary Yoneda, it suffices to show that $\eta_C 1$ is a fixed point and that η defined by $\eta_D\gamma = (F\gamma)x$, $\gamma \in \mathcal{C}(C,D)$, $x \in ({}^{G}F)C$, is G-natural. For the former

$${}^{g}(\eta_C 1) = ({}^{g}\eta_C){}^{g}1 = \eta_{{}^{g}C}{}^{g}1 = \eta_C 1$$

and, for the latter

$$({}^{g}\eta_D){}^{g}\gamma = {}^{g}(\eta_D\gamma) = {}^{g}((F\gamma)x) = {}^{g}(F\gamma){}^{g}x = (F{}^{g}\gamma)x = \eta_{{}^{g}D}{}^{g}\gamma. \quad \square$$

Using this, the nature of G-representability may be elucidated as in

COROLLARY 4.5. A G-functor $F\colon \mathcal{C} \to \mathrm{par}(G\text{-Set})$ is G-representable by $C \in \mathrm{stab}\,\mathcal{C}$ precisely when there exists an obj(stab $\mathcal{C}$)-indexed family of equivariant isomorphisms $\omega_D\colon \mathcal{C}(C,D) \to FD$ natural in D, in which case the ω_D constitute the Dth components of a unique G-representation of F by C.

Proof. By equivariance of ω_C, $x = \omega_C 1$ is a fixed point of FC. Hence there is a unique G-natural transformation $\eta\colon \mathcal{C}(C,-) \to F$ such that $\eta_C 1 = x$. But, by naturality of the ω_D in D, $\omega_D\gamma = (F\gamma)x$ for all $\gamma \in \mathcal{C}(C,D)$. Thus $\eta_D = \omega_D$ so that, in particular, the η_D are isomorphisms. $\square$

An easy consequence is that every G-representation of F is a stable representation provided objects D of $\mathcal{C}$ such that FD is nonempty are isomorphic to stable objects.

Notice that Yoneda for par(G-Set)-valued H-functors is an automatic consequence of Lemma 4.4; that is

$$[\mathrm{stab}_H(\mathrm{par}(G\text{-Set}))^{\mathcal{C}}](\mathcal{C}(C,-),F) \cong ({}^{H}F)C, \qquad C \in \mathrm{stab}_H\mathcal{C}$$

by the Yoneda map (inasmuch as the left hand side is canonically isomorphic to $[\mathrm{stab}(\mathrm{par}(H\text{-Set}))^{\mathcal{C}}](\mathcal{C}(C,-),UF)$). In particular, applying Lemma 3.1 one has

THEOREM 4.6 (cf. 6.13). A G-functor N: $\mathcal{Q} \to \mathcal{P}$ has a left G-adjoint when and only when, for each $P \varepsilon \mathcal{P}$, the G(P)-functor $\mathcal{P}(P,N-)$: $\mathcal{Q} \to$ par(G-Set) is G(P)-stably representable. □

Returning to the previous notation:

THEREOM 4.7. Given a stable object C of $\mathcal{C}$

$$\begin{array}{ccc} [\mathrm{stab}(\mathrm{par}(G\text{-Set}))^{\mathcal{C}}](\mathcal{C}(C,-),F) & \xrightarrow{G_{-}} & \mathrm{Set}^{\mathrm{stab}\ \mathcal{C}}([\mathrm{stab}\ \mathcal{C}](C,-),{}^{G}F) \\ & \searrow \quad ({}^{G}F)C \quad \swarrow & \end{array}$$

is a commutative triangle of natural isomorphisms, the legs being Yoneda maps. Moreover, if η is a G-representation of F by C, ${}^{G}\eta$ is a represenatation of ${}^{G}F$ by C.

Proof. Since it is plain the triangle commutes and since, by 4.4 and ordinary Yoneda, the legs are isomorphisms, ${}^{G}-$ pictured is an isomorphism. The statement concerning G-representations is clear -- ${}^{G}\eta$ at stable $D \varepsilon \mathcal{C}$ is the restriction of the isomorphism η_D to stab$\mathcal{C}(C,D)$. □

COROLLARY 4.8. If N: $\mathcal{Q} \to \mathcal{P}$ has a left G-adjoint then stab N: stab $\mathcal{Q} \to$ stab $\mathcal{P}$ has a left adjoint. □

Concerning Thereom 4.7, we hasten to mention that not every representation of ${}^{G}F$ need come from a G-representation of F. However, at least for the example of cones to a G-diagram in $\mathcal{C}$ -- which we discuss next -- the contrary case has an important characterization (see 4.10).

The contravariant functor cone(−,Z), Z a G-diagram in $\mathcal{C}$ over J, as the composite

$$\mathcal{C}^{op} \xrightarrow{\Delta^{op}} (\dot{\mathcal{C}}^{J})^{op} \xrightarrow{\dot{\mathcal{C}}^{J}(-,Z)} \mathrm{par}(G\text{-Set})$$

is a G-functor and, evidently, ^{G}[cone(−,Z)] = G-cone(−,Z). Now, under the Yoneda map of Lemma 4.4, stable representations and G-representations of cone(−,Z) correspond, respectively, to stable limits of Z and to G-cones to Z through which every cone from a stable object to Z factors uniquely. Theorem 4.7 implies that a G-cone to Z of the latter type, κ say, is already a limiting

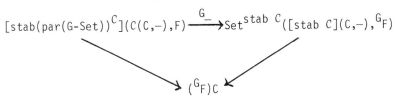

G-cone -- cf. Lemma 1.2. Thus, if every G-limit of Z is a limit or, perchance, specializing the comment following Corollary 4.5, every object admitting a cone to Z is isomorphic to a stable object, κ will be a limit (thence corresponding to a stable representation).

Using Theorem 4.3, we may easily construct all H-stable limits in par(G-Set):

PROPOSITION 4.9. Let Σ be an H-diagram in par(G-Set) over J, and denote the one point set by $*$. Then H-cone$(H\cdot *,\Sigma)$ carries a unique H-set structure for which the cone σ: H-cone$(H\cdot *,\Sigma) \to \Sigma$ defined by

$$\sigma_j\chi = \chi_j^*, \qquad \chi \in \text{H-cone}(H\cdot *,\Sigma),\ j\varepsilon J$$

is an H-cone; and this σ affords an H-stable limit for Σ.

Proof. Choose an H-set X and H-stable limit Θ: $X \to \Sigma$ and let σ be the limiting cone defined by commutativity of the diagram

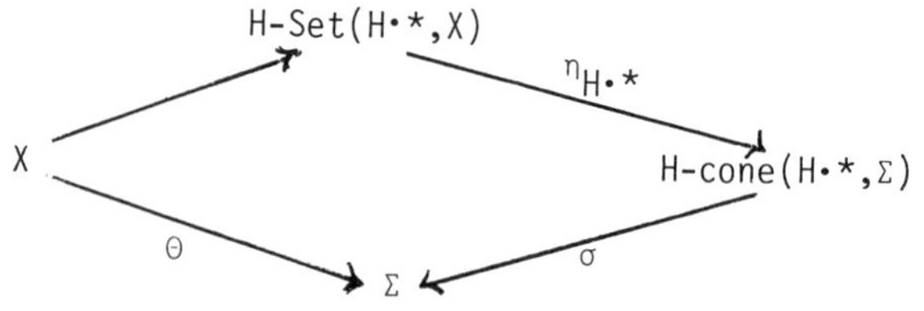

where the unlabeled arrow is the bijection given by the universal property of $H\cdot *$ and η is the representation of H-cone$(-,\Sigma)$ corresponding to the H-limit Θ. □

Explicitly, the action of H on H-cone$(H\cdot *,\Sigma)$ is given by $(h\chi)_j{}^{k*} = \chi_j{}^{kh*}$. We use Proposition 4.9 in the following manner. Suppose that Θ: $\lim_G Z \to Z$ is a G-limit, where Z is a G-diagram in C over J. Let, using Theorem 4.7, η: $C(-,\lim_G Z) \to \text{cone}(-,Z)$ be the G-natural transformation such that ${}^G\eta$ is the representation of G-cone$(-,Z)$ corresponding to Θ (i.e., $\eta_{\lim_G Z}1 = \Theta$). Then, for $C \in \text{stab}\, C$

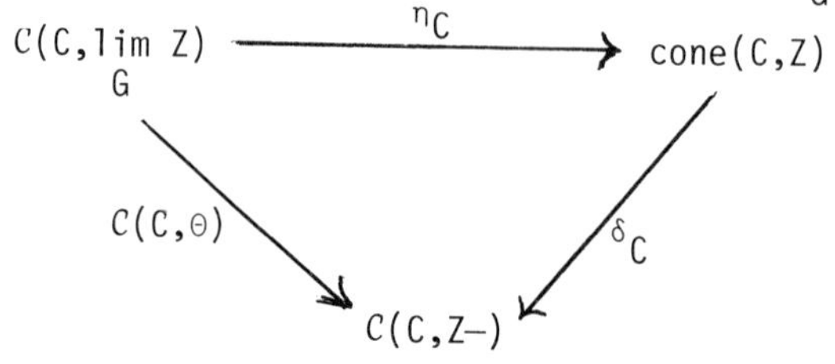

is a commutative diagram of G-natural transformations of par(G-Set)-valued G-functors of J (only $C(C,Z-)$ is nonconstant), natural in C, where δ_C is defined on cones γ: $C \to Z$ by $(\delta_C)_j\gamma = \gamma_j$, $j \in J$. But, plainly, δ_C is the composite

$$\text{cone}(C,Z) \xrightarrow{f_C} \text{G-cone}(G\cdot *, C(C,Z-)) \xrightarrow{\sigma_C} C(C,Z-)$$

where σ_C is the limiting G-cone of Proposition 4.9 and f_C is the (necessarily G-equivariant) isomorphism given by $(f_C\gamma)_j{}^{g*} = (g\cdot\gamma)_j$. In other words, δ_C is a limiting G-cone; and thus commutativity of the above diagram shows:

THEOREM 4.10. Given a G-diagram Z in C, the functor $^G-$ induces a bijection between G-representations of cone(–,Z) and representations of G-cone(–,Z) if and only if, for all C ε stab C, the par(G-Set)-valued G-hom-functor $C(C,-)$ preserves G-limits of Z. □

This result, put differently, asserts that $C(C,-)$ preserves G-limits of Z for all stable C exactly when all cones from a stable object to Z have a unique factorization through every G-limit of Z. It follows that when, for example, stab C is a replete subcategory of C, every G-limit in C is a limit if and only if the $C(C,-)$ are G-continuous for all stable C. Clearly, this is also a consequence of Theorem 4.3. The next result is more precise.

COROLLARY 4.11. All G-limits of Z are limits if and only if $C(C,-)$ preserves G-limits of Z for every stable C and all G-representations of cone(–,Z) are stable representations. □

Despite the fact that G-hom-functors are not necessarily G-continuous, as indicated in the preceding section, right G-adjoints necessarily are G-continuous. We could, at this point, handily give a G-representation theoretic proof thereof. However, we defer this to the section after next, where more general results are proved for "transversaled adjoints" -- results analogous to G-continuity for transversaled adjoints can no longer be claimed to be easy computations.

5. TRANSVERSALS

A *transversal* for an object of a $G \times G^{op}$-category X is a G-indexed family $\tau = \{\tau_g\}$ of isomorphisms $\tau_g\colon {}^gX \to X^g$ such that the triangle

$$\begin{array}{ccc} {}^{gh}X & \xrightarrow{\;{}^g\tau_h\;} & {}^gX^h \\ & {\scriptstyle\tau_{gh}}\searrow \quad \swarrow{\scriptstyle\tau_g^h} & \\ & X^{gh} & \end{array} \qquad (1)$$

commutes for all $g,h \in G$. We say that X is *transversaled by* τ and call the pair (X,τ) a *transveraled object of* X. When, for all g, ${}^gX = X^g$ and $\tau_g = 1$ we use the term *trivially transversaled*. A *transverse arrow* $\xi\colon (X,\tau) \to (X',\tau')$ is an arrow $\xi\colon X \to X'$ in X such that

$$\begin{array}{ccc} {}^gX & \xrightarrow{\;{}^g\xi\;} & {}^gX' \\ {\scriptstyle\tau_g}\downarrow & & \downarrow{\scriptstyle\tau'_g} \\ X^g & \xrightarrow[\;\xi^g\;]{} & X'^g \end{array}$$

commutes for all g. The transversaled objects and transverse arrows form a category, trans X, compositions and identities being those of X. We denote the full subcategory of trivially transversaled objects by triv-trans X. The category, cotrans X, of *cotransversaled* objects and *cotransverse* arrows of X is defined by $(\text{cotrans } X)^{op} = \text{trans } X^{op}$.

In that any G-category is a $G \times G^{op}$-catgeory with the trivial action of G^{op}, a transversaled object in a G-category is the same thing as a system of isomorphisms. More generally, if (X,τ) is a transversaled object of X, since

$$\tau_g^{g^{-1}} \circ {}^g\tau_h^{h^{-1}g^{-1}} = (\tau_g^h \circ {}^g\tau_h)^{h^{-1}g^{-1}} = \tau_{gh}^{(gh)^{-1}}$$

$\dot\tau$ defined by $\dot\tau_g = \tau_g^{g^{-1}}$ is a system of isomorphisms at X in $\dot X$. It follows that the assignments

$$(X,\tau) \longmapsto (X,\dot\tau) \qquad\qquad \xi \longmapsto \xi$$

yield an isomorphism trans $X \to$ trans $\dot X$ = stab $\dot X^G$. In particular, arrows

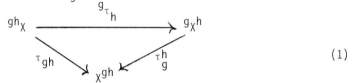

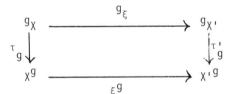

τ_g making the diagram (1) commute constitute a transversal if and only $\tau_1 = 1$ and, if $\tau = \{\tau_g\}$ is a transversal, $\tau^{-1} = \{\tau_g^{-1}\}$ is a cotransversal ($\tau_g^{-1} = {}^g\tau_{g^{-1}}$). Also, notice that a transversaled object is transverse isomorphic to a trivially transversaled object if and only if the corresponding system of isomorphisms is stabilizing. We call such transversaled objects *trivializable* and their transversals *trivializing*.

If A and $\mathcal{B}$ are G-categories, we call objects of trans $\mathcal{B}^A$ *transversaled functors* and arrows *transverse natural transformations*. Note that to say a transversaled functor (T,τ) is trivializable is to say it is "trivialized" by a G-functor S -- or, more precisely, there is a natural isomorphism $\varepsilon\colon S \to T$ such that $\tau_g = \varepsilon^g \cdot {}^g\varepsilon^{-1}$ for every g in G. In fact, by Corollary 2.4, a functor is naturally isomorphic to a G-functor exactly when it has a trivializing transversal. Heavy use of transversaled functors is made in [14]. Given trivializability, the theory reverts to the cleaner, more intuitive theory of G-functors. In this sense Theorem 5.7 below, asserting transversaled functors between hereditarily stably closed G-categories are trivializable provided they have adjoints, is the major result of the section. Yet, from another viewpoint, as we shall see the very generality of transversaled functors serves to clarify, unify and extend previous notions and results.

We take composition of transversaled functors to be given by

$$(T',\tau')\circ(\Gamma,\tau) = (T'T,\tau'\tau), \qquad (\tau'\tau)_g = T'\tau_g \cdot \tau'_gT. \tag{2}$$

Writing $\tau'_gT^h \cdot {}^gT'\tau_h$ and $T'{}^g\tau_g \cdot \tau'^h_gT$ as vertical compositions of horizontal composites and using the interchange law, we find both expressions to be equal to the same horizontal composite $\tau'_g \circ \tau_h$. Thus

$$(\tau'\tau)^h_g \cdot {}^g(\tau'\tau)_h = T'\tau^h_g \cdot \tau'_gT^h \cdot {}^gT'\tau_h \cdot {}^g\tau'_hT = T'\tau^h_g \cdot T'{}^g\tau_h \cdot \tau'^h_gT \cdot {}^g\tau'_hT$$
$$= T'(\tau^h_g \cdot {}^g\tau_h) \cdot (\tau'^h_g \cdot {}^g\tau'_h)T = T'\tau_{gh} \cdot \tau'_{gh}T = (\tau'\tau)_{gh}$$

and, consequently, the composition (2) is defined whenever $T'T$ is defined. With this composition, we form the category Trans-Cat of G-categories and transversaled functors.

Now, if η: $(M,\mu) \to (N,\nu)$: $A \to B$ is a transverse natural transformation and (T,τ): $B \to C$ and (S,σ): $D \to A$ are transversaled functors, the diagrams

$$\begin{array}{ccc} {}^{g}TM & \xrightarrow{\ {}^{g}T\eta\ } & {}^{g}TN \\ \downarrow{\scriptstyle \tau_g{}^{M}} & & \downarrow{\scriptstyle \tau_g{}^{N}} \\ T{}^{g}M & \xrightarrow{\ T{}^{g}\eta\ } & T{}^{g}N \\ \downarrow{\scriptstyle T\mu_g} & & \downarrow{\scriptstyle T\nu_g} \\ TM^{g} & \xrightarrow{\ T\eta^{g}\ } & TN^{g} \end{array} \qquad\qquad \begin{array}{ccc} {}^{g}MS & \xrightarrow{\ {}^{g}\eta S\ } & {}^{g}NS \\ \downarrow{\scriptstyle \mu_g{}^{S}} & & \downarrow{\scriptstyle \nu_g{}^{S}} \\ M{}^{g}S & \xrightarrow{\ \eta{}^{g}S\ } & N{}^{g}S \\ \downarrow{\scriptstyle M\sigma_g} & & \downarrow{\scriptstyle N\sigma_g} \\ MS^{g} & \xrightarrow{\ \eta S^{g}\ } & NS^{g} \end{array}$$

are commutative, because the upper square on the left commutes by naturality of τ_g and the lower square on the right commutes by naturality of η. Hence $T\eta$: $(TM,\tau\mu) \to (TN,\tau\nu)$ and ηS: $(MS,\mu\sigma) \to (NS,\nu\sigma)$ are transverse natural transformations. It follows that Trans-Cat is a 2-category under the usual horizontal and vertical composition of Cat. We denote the sub-2-category defined by the trivially transversaled functors by G-Cat. In other words, G-Cat is the category of G-categories, G-functors, and G-natural transformations, the last being the 2-cells. Additionally, we see that an adjunction in this 2-category is exactly a G-adjunction as defined previously -- see, for example, [20, pp. 84 and 85].

We define a *transversaled adjunction* to be an adjunction in the 2-category Trans-Cat. Thus a transversaled adjunction consists of 1-cells (M,μ): $P \to Q$ and (N,ν): $Q \to P$ and 2-cells λ: $1 \to (NM,\nu\mu)$ and ρ: $(MN,\mu\nu) \to 1$ such that the usual triangle identities for λ and ρ are valid. That λ and ρ are 2-cells means they are natural transformations for which the triangles

$$\begin{array}{ccc} & {}^{g}\_ & \\ {\scriptstyle \lambda^{g}}\swarrow & & \searrow{\scriptstyle {}^{g}\lambda} \\ NM^{g} & \xleftarrow[\ N\mu_g \,\cdot\, \nu_g{}^{M}\ ]{} & {}^{g}NM \end{array} \;. \qquad\qquad \begin{array}{ccc} {}^{g}MN & \xrightarrow{\ M\nu_g \,\cdot\, \mu_g{}^{N}\ } & MN^{g} \\ {\scriptstyle {}^{g}\rho}\searrow & & \swarrow{\scriptstyle \rho^{g}} \\ & {}^{g}\_ & \end{array} \qquad (3)$$

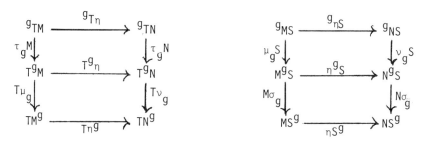

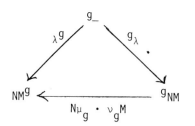

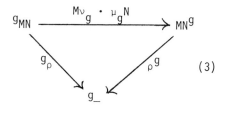

commute. Note that, from commutativity of the left triangle, ν_g determines μ_g and, from commutativity of the right triangle, μ_g determines ν_g. As with G-adjunctions, in the presence of the triangle identities, λ is a 2-cell if and only if ρ is a 2-cell. For, we maintain, the adjunction isomorphism $\Psi = \mathcal{P}(\lambda,1) \cdot N\colon \mathcal{Q}(M-,-) \to \mathcal{P}(-,N-)$ is a 2-cell precisely when λ is a 2-cell. But, it is readily checked, using naturality of ν_g, that $N\colon\mathcal{Q}(M-,-) \to \mathcal{P}(NM-,N-)$ is a 2-cell ; that is

$$\begin{array}{ccc}
\mathcal{Q}({}^gM-,{}^g-) & \xrightarrow{\ {}^gN\ } & \mathcal{P}({}^gNM-,{}^gN-) \\
{\scriptstyle \mathcal{Q}(\mu_g^{-1},1)}\big\downarrow & & \big\downarrow{\scriptstyle \mathcal{P}((\nu\mu)_g^{-1},\nu_g)} \\
\mathcal{Q}(M^g-,{}^g-) & \xrightarrow[\ N^g\ ]{} & \mathcal{P}(NM^g,N^g-)
\end{array}$$

commutes. Thus we see that Ψ is a 2-cell -- meaning

$$\begin{array}{ccc}
\mathcal{Q}({}^gM-,{}^g-) & \xrightarrow{\ {}^g\Psi\ } & \mathcal{P}({}^g-,{}^gN-) \\
{\scriptstyle \mathcal{Q}(\mu_g^{-1},1)}\big\downarrow & & \big\downarrow{\scriptstyle \mathcal{P}(1,\nu_g)} \\
\mathcal{Q}({}^gM-,{}^g-) & \xrightarrow[\ \Psi^g\ ]{} & \mathcal{P}({}^g-,N^g-)
\end{array}$$

is commutative -- if and only if

$$\mathcal{P}({}^g\lambda,\nu_g) \cdot {}^gN = \mathcal{P}(\lambda^g \cdot (\nu\mu)_g^{-1},\nu_g) \cdot {}^gN.$$

This, by Yoneda, is equivalent to λ being a 2-cell. In particular, we have shown:

PROPOSITION 5.1. If (M,μ) and (N,ν) are transversaled functors and M is left adjoint to N with unit a transverse natural transformation, then (M,μ) is left transverse-adjoint to (N,ν). □

Retaining the same notation, given the transversal ν for N, define $\bar{\nu}_g\colon M^g \to {}^gM$ to be the vertical composite

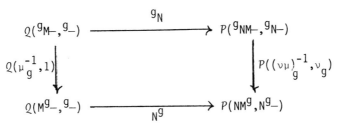

$$M^g \xrightarrow{M^g\lambda} M^gNM \xrightarrow{M\nu_g M} MN^gM \xrightarrow{\rho\, {}^gM} {}^gM \;.$$

Since in the diagram

$$\begin{array}{ccccccc}
{}^g_ & \xrightarrow{{}^g\lambda} & {}^gNM & \xrightarrow{\nu_g M} & N^gM & & \\
\downarrow{\scriptstyle \lambda^g} & & \downarrow{\scriptstyle \lambda^gNM} & & \downarrow{\scriptstyle \lambda N^gM} & \diagdown\!\!\diagdown & \\
NM^g & \xrightarrow[NM^g\lambda]{} & NM^gNM & \xrightarrow[NM\nu_g M]{} & NMN^gM & \xrightarrow[N\rho\, {}^gM]{} & N^gM
\end{array}$$

the squares commute by naturality of λ and the triangle is one of the triangle identities, $\bar{\nu}_g$ is equally defined by commutativity of

$$\begin{array}{ccc}
{}^g_ & \xrightarrow{{}^g\lambda} & {}^gNM \\
\downarrow{\scriptstyle \lambda^g} & & \downarrow{\scriptstyle \nu_g M} \\
NM^g & \xrightarrow[N\bar{\nu}_g]{} & N^gM
\end{array} \;. \qquad (4)$$

In particular, $\bar{\nu}_g$ is an isomorphism (since $\nu_g M$ is) and, since

$$\begin{array}{ccccc}
 & & {}^{gh}_ & \xrightarrow{{}^{gh}\lambda} & {}^{gh}NM \\
 & \diagup\!\!\diagup & \downarrow{\scriptstyle {}^g\lambda^h} & & \downarrow{\scriptstyle {}^g\nu_h M} \\
{}^{gh}_ & \xrightarrow{{}^g\lambda^h} & {}^gNM^h & \xrightarrow{{}^gN\bar{\nu}_h} & {}^gN^hM \\
\downarrow{\scriptstyle \lambda^{gh}} & & \downarrow{\scriptstyle \nu_g M^h} & & \downarrow{\scriptstyle \nu_g {}^hM} \\
NM^{gh} & \xrightarrow[N\bar{\nu}_g^h]{} & N^gM^h & \xrightarrow[N^g\bar{\nu}_h]{} & N^{gh}M
\end{array}$$

commutes, ${}^g\bar{\nu}_h \cdot \bar{\nu}_g^h = \bar{\nu}_{gh}$. Thus the $\bar{\nu}_g$ form a cotransversal for M. We call $\bar{\nu} = \{\bar{\nu}_g\}$ the *companion cotransversal of* ν.

All told, we have proved

PROPOSITION 5.2. If M is a left adjoint of N and N is tranversaled by ν, there is a unique transversal μ for M such that (M,μ) is a left transversaled adjoint of (N,ν) (with the same unit); namely, $\mu = \bar{\nu}^{-1}$. □

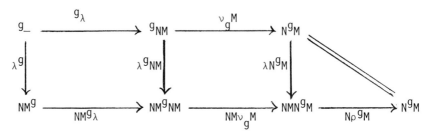

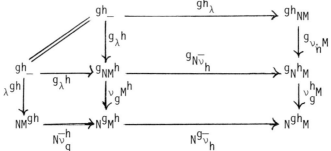

THEOREM 5.3. Given a left adjoint M of N, the map $\nu \longmapsto \bar{\nu}$ taking transversals for N to their companion cotransversals is a bijection between transversals for N and cotransversals for M.

Proof. By duality, it suffices to show that $\bar{\bar{\nu}} = \nu$. Now, as we saw above, $\bar{\bar{\nu}}$ is defined by commutativity of the square

$$\begin{array}{ccc} M^gN & \xrightarrow{\;M\bar{\bar{\nu}}_g\;} & MN^g \\ {\scriptstyle \bar{\nu}_g N}\downarrow & & \downarrow{\scriptstyle \rho^g} \\ {}^gMN & \xrightarrow[\;{}^g\rho\;]{} & {}^g\_ \end{array} .$$

But $(M,\bar{\nu}^{-1})$ is a 1- cell and, inasmuch as (4) commutes, λ is a 2-cell. So, by 5.1, ρ is a 2- cell or, otherwise put, the square just displayed commutes with $\bar{\bar{\nu}}_g$ replaced by ν_g. □

We should comment that the definition of $\bar{\nu}$ itself affords a more direct proof of the equivalence of the commutative triangles in (3) (and hence of 5.1) than the proof given. Indeed, this assertion is an immediate consequence of the fact that, by naturality of ν_g and ρ^g

$$\begin{aligned} \rho^g \cdot M\nu_g &= \rho^g \cdot MN^g\rho \cdot M\nu_g MN \cdot M^g\lambda N = {}^g\rho \cdot \rho^g MN \cdot M\nu_g MN \cdot M^g\lambda N \\ &= {}^g\rho \cdot (\rho^g M \cdot M\nu_g M \cdot M^g\lambda)N = {}^g\rho \cdot \bar{\nu}_g N \end{aligned}$$

so that, again, $\bar{\bar{\nu}} = \nu$.

COROLLARY 5.4. If (M,μ) is left transverse-adjoint to (N,ν) with unit λ and counit ρ then the following are equivalent.

(i) μ is trivial;

(ii)

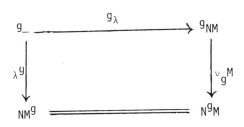

$$\begin{array}{ccc} {}^g\_ & \xrightarrow{\;{}^g\lambda\;} & {}^gNM \\ {\scriptstyle \lambda^g}\downarrow & & \downarrow{\scriptstyle \nu_g M} \\ NM^g & =\!= & N^gM \end{array}$$

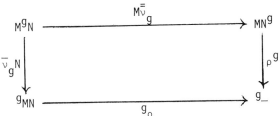

is a commutative square;

(iii) M is a G-functor and $\nu_g = N^g\rho \cdot \lambda^g N$;

(iv) M is a G-functor and $\rho^g \cdot M\nu_g = {}^g\rho$. □

To proceed to the generalization of Lemma 3.1 promised in Section 3 one must observe that, for $P \varepsilon \mathcal{P}$ transversaled, say, by ϕ (i.e., ϕ is a system of isomorphisms at P), $(P{\downarrow}N)$ is a G-category: defining

$$ {}^g(Q,p) = ({}^gQ, \nu_g(Q)\,{}^gp\phi_g^{-1}), \quad Q\varepsilon\mathcal{Q},\ p\varepsilon\mathcal{P}(P,NQ) $$

one has

$$ {}^h({}^g(Q,p)) = {}^h({}^gQ, \nu_g(Q)\,{}^gp\phi_g^{-1}) = ({}^{hg}Q, \nu_h({}^gQ)\,{}^h(\nu_g(Q)\,{}^gp\phi_g^{-1})\phi_h^{-1}) $$

$$ = ({}^{hg}Q, \nu_h^g(Q)\,{}^h\nu_g(Q)\,{}^{hg}p\,{}^h\phi_g^{-1}\phi_h^{-1}) = ({}^{hg}Q, \nu_{hg}(Q)\,{}^{hg}p\phi_{hg}^{-1}) $$

$$ = {}^{hg}(Q,p) $$

and obviously, ${}^1(Q,p) = (Q,p)$. Thereto, by naturality of the ν_g, the operation of G on arr$\mathcal{Q}$ does indeed yield an operation of G on arr$(P{\downarrow}N)$. Notice that, taking ϕ to be the trivial G(P)-transversal for P, one gets that $(P{\downarrow}N)$ is a G(P)-category. By Proposition 5.2, this G(P)-category has an initial object for all P precisely when (N,ν) has a left transversaled adjoint.

LEMMA 5.5. The transversaled functor $(N,\nu)\colon \mathcal{Q} \to \mathcal{P}$ has a left trivially transversaled adjoint if and only if, for all $P\varepsilon\mathcal{P}$, the G(P)-category $(P{\downarrow}N)$ has a stable initial object.

Proof. $\Longrightarrow$. If M is a left trivially transversaled adjoint of (N,ν) with unit λ then, by 5.4, (MP,λ_P) is the required stable initial object.

$\Longleftarrow$. Choose representative objects R of $\mathcal{P}$, one in each orbit under the action of G, and let, say, (MR,λ_R) be a G(R)-stable initial object of $(R{\downarrow}N)$. So, for all $k \varepsilon G(R)$, ${}^kMR = MR$ and

$$ \begin{array}{ccc} R & \xrightarrow{\ {}^k\lambda_R\ } & {}^kNMR \\ {\scriptstyle\lambda_R}\downarrow & & \downarrow{\scriptstyle\nu_k(MR)} \\ NMR & = & N^kMR \end{array} $$

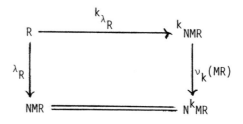

commutes. Now, if $P \varepsilon \mathcal{P}$, $P = {}^hR$ for unique R and some $h \varepsilon G$. Define

$$MP = {}^hMR \qquad \lambda_P = \nu_h(MR)^h\lambda_R \quad .$$

Clearly, MP is well defined. As for λ_P, suppose ${}^hR = {}^gR$, $g \varepsilon G$. Since replacing k in the displayed commutative diagram with $g^{-1}h$ and composing with ${}^g\_$, one sees that

$$\nu_g(MR)^g\nu_{g^{-1}h}(MR) = \nu_g^{g^{-1}h}(MR)^g\nu_{g^{-1}h}(MR) = \nu_h(MR),$$

one gets the commutative diagram

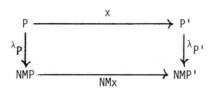

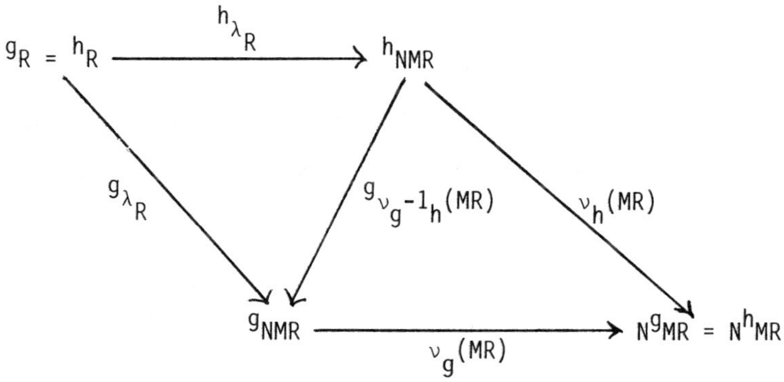

Thus λ_P is well defined.

Next, notice one has $M^gP = {}^{gh}MR = {}^gMP$. For $\lambda_{{}^gP}$, one computes:

$$\begin{aligned}\lambda_{{}^gP} &= \nu_{gh}(MR)^{gh}\lambda_R = \nu_{gh}(MR)^g(\nu_h^{-1}(MR)\lambda_P) \\ &= \nu_{gh}(MR)^g\nu_h^{-1}(MR)^g\lambda_P = \nu_{gh}(MR)^{gh}\nu_{h^{-1}}(MR)^g\lambda_P \\ &= \nu_{gh}^{h^{-1}}(MP)^{gh}\nu_{h^{-1}}(MP)^g\lambda_P = \nu_g(MP)^g\lambda_P.\end{aligned}$$

Thus, provided M can be defined on arrows so as to be a G-functor, λ is a 2-cell. But, inasmuch as $(MP,\lambda_P) = {}^h(MR,\lambda_R)$ is initial, Mx must be defined for $x \varepsilon \mathcal{P}(P,P')$ by commutativity of

$$\begin{array}{ccc} P & \xrightarrow{\quad x \quad} & P' \\ {\scriptstyle \lambda_P}\downarrow & & \downarrow{\scriptstyle \lambda_{P'}} \\ NMP & \xrightarrow[\quad NMx \quad]{} & NMP' \end{array} \quad .$$

But then, it follows that

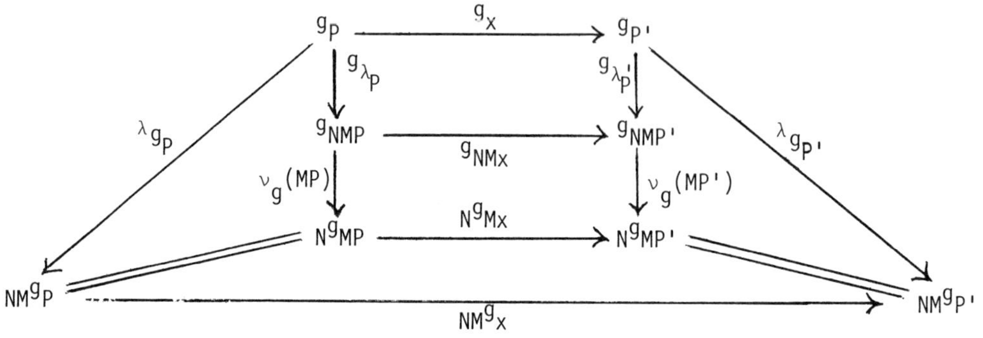

commutes; and so ${}^{g}Mx = M^{g}x$.

In the light of 5.1, all that is needed to finish is as in the usual pointwise construction of adjoints: ρ defined by commutativity of

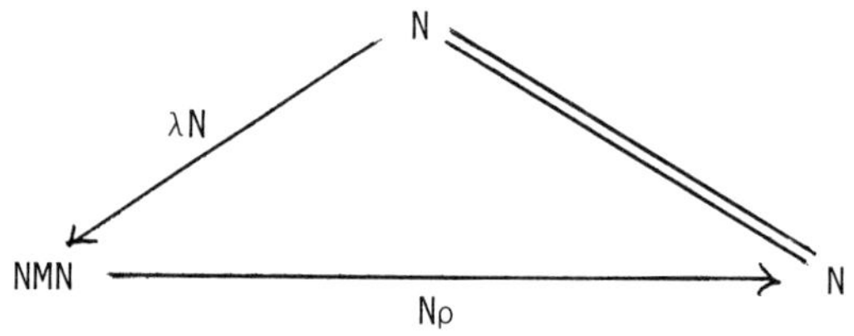

makes, by naturality of λ

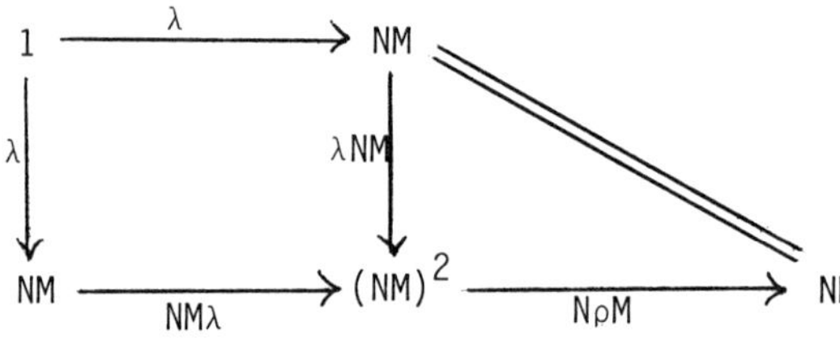

commute; so $\rho M \cdot M\lambda = 1$. □

Now

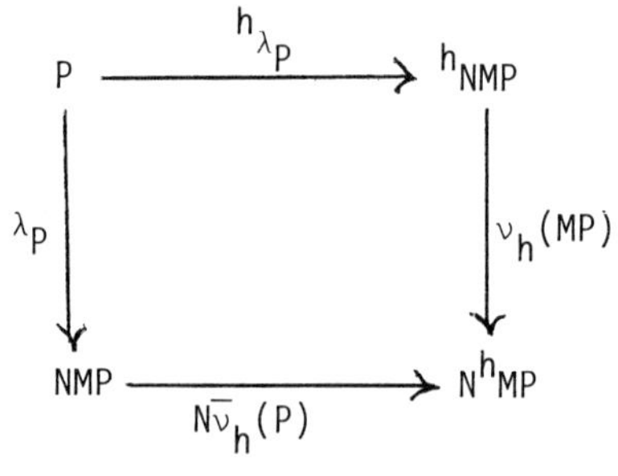

is a commutative diagram for all $h \in G(P)$. Thus, by the properties of the projection $(P{\downarrow}N) \to \mathcal{Q}$, Corollary 2.5 and Propositions 1.7 and 2.9, we get the following extension of Theorem 3.10.

THEOREM 5.6. A transversaled functor $(N,\nu)\colon \mathcal{Q} \to \mathcal{P}$ with left adjoint M has a left trivially transversaled adjoint if and only if, for all $P \in \mathcal{P}$, the G(P)-cosystem of isomorphisms in $\mathcal{Q}$

$$\{\bar{\nu}_h(P)\colon MP \to {}^hMP\}_{h \in G(P)}$$

is costabilizing, where $\bar{\nu}$ is the companion cotransversal of ν. □

If (M,1) is a left transversaled adjoint of (N,ν), plainly, (N,ν) is trivializable precisely when M has a right G-adjoint. So we have

THEOREM 5.7. If a transversaled functor defined on an hereditarily stably closed G-category has a left adjoint, then it has a left trivially transversaled adjoint. Thus, a transversaled functor between hereditarily stably closed G-categories is trivializable whenever it has a left adjoint. □

This result is best possible in the sense that a transversaled functor with hereditarily stably closed domain can have a left adjoint and yet not be trivializable.

6. TRANSVERSE LIMITS AND REPRESENTATIONS OF TRANSVERSALED FUNCTORS

Let $\mathcal{X}$ and $\mathcal{Y}$ be $G \times G^{op}$-categories, let $F: \mathcal{X} \to \mathcal{Y}$ be a G^{op}-functor, and let $\delta = \{\delta_g\}$ be a G-indexed family of G^{op}-natural isomorphisms $\delta_g: {}^gF \to F^g$ subject to $\delta_g^h \cdot {}^g\delta_h = \delta_{gh}$ for all $g,h \varepsilon G$. Now, as is readily seen, these equations mean that the family $\dot{\delta} = \{\dot{\delta}_g\}$, where $\dot{\delta}_g(X) = \delta_g(X^{g^{-1}})$, is a transversal for F considered as functor $\dot{\mathcal{X}} \to \dot{\mathcal{Y}}$. In particular, viewing transversaled objects of $\dot{\mathcal{X}}$ and $\dot{\mathcal{Y}}$ as transversaled functors of the one-object discrete (trivial) G-category, we get a functor trans $\dot{\mathcal{X}} \to$ trans $\dot{\mathcal{Y}}$ as described by

$$(X,\tau) \longmapsto (FX,\dot{\delta}\tau), \qquad (\dot{\delta}\tau)_g = F\tau_g \circ \dot{\delta}_g(X).$$

We define trans F: trans $\mathcal{X} \to$ trans $\mathcal{Y}$ to be the composite functor

$$\text{trans } \mathcal{X} \to \text{trans } \dot{\mathcal{X}} \to \text{trans } \dot{\mathcal{Y}} \to \text{trans } \mathcal{Y}.$$

Thus, on objects, $\text{trans}F(X,\tau) = (FX,\delta\tau)$, where $(\delta\tau)_g = F\tau_g \circ \delta_g(X)$, and where, of course, on arrows ξ $(\text{trans } F)\xi = F\xi$. We denote the restriction of trans F to triv-trans $\mathcal{X}$ by triv-trans F. Notice that if $E: \mathcal{X} \to \mathcal{Y}$ is a $G \times G^{op}$-functor then, as functor $\dot{\mathcal{X}} \to \dot{\mathcal{Y}}$, E is trivially transversaled; and trans E is the evident functor.

The following result will play only a minor role in the section, and is chiefly designed for use in [14].

PROPOSITION 6.1. In the above notation, suppose $\pi: (Y,t) \to \text{trans}F(X,\tau)$ is a transverse arrow in $\mathcal{Y}$ such that (X,π) is initial in the comma category $(Y{\downarrow}F)$. Then $((X,\tau),\pi)$ is an initial object of $((Y,t){\downarrow}\text{trans } F)$.

Proof. Without loss of generality, $\mathcal{X}$ and $\mathcal{Y}$ are trivial G^{op}-categories. We claim that $({}^gX, \delta_g(X) \circ {}^g\pi)$ is an initial object of $({}^gY{\downarrow}F)$: Let $a \in \mathcal{Y}({}^gY, FX')$. Then

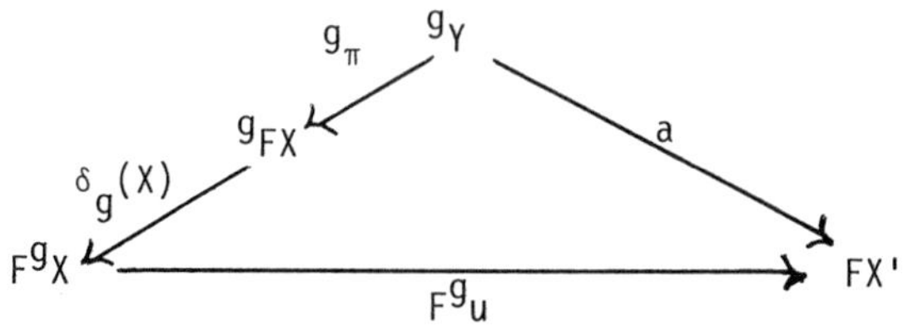

(1)

commutes for the $u \in X(X, {}^{g^{-1}}X')$ making

$$\begin{array}{ccc}
Y & \xrightarrow{\;{}^{g^{-1}}a\;} & {}^{g^{-1}}FX' \\
{\scriptstyle\pi}\big\downarrow & & \big\downarrow{\scriptstyle\delta_{g^{-1}}(X')} \\
FX & \xrightarrow[\;Fu\;]{} & F^{g^{-1}}X'
\end{array}$$

commute since, by naturality of δ_g

$$\begin{array}{ccc}
{}^{g}Y & \xrightarrow{\;a\;} & FX' \\
{\scriptstyle {}^{g}\pi}\big\downarrow & & \big\downarrow{\scriptstyle {}^{g}\delta_{g^{-1}}(X')} \\
{}^{g}FX & \xrightarrow{\;{}^{g}Fu\;} & {}^{g}F^{g^{-1}}X' \\
{\scriptstyle \delta_g(X)}\big\downarrow & & \big\downarrow{\scriptstyle \delta_g({}^{g^{-1}}X') \,=\, \delta_g^{g^{-1}}(X')} \\
F^{g}X & \xrightarrow[\;F^{g}u\;]{} & FX'
\end{array}$$

commutes. Also, if (1) commutes with ${}^{g}u$ replaced by v

$$\begin{array}{ccccc}
 & & Y & & \\
 & {\scriptstyle\pi}\swarrow & & \searrow{\scriptstyle {}^{g^{-1}}a} & \\
FX & & & & \\
{\scriptstyle {}^{g^{-1}}\delta_g(X)}\swarrow & & & & \\
{}^{g^{-1}}F^{g}X & & \xrightarrow{\;{}^{g^{-1}}Fv\;} & & {}^{g^{-1}}FX' \\
{\scriptstyle \delta_{g^{-1}}({}^{g}X)}\big\downarrow & & & & \big\downarrow{\scriptstyle \delta_{g^{-1}}(X')} \\
FX & & \xrightarrow[\;F^{g^{-1}}v\;]{} & & F^{g^{-1}}X'
\end{array}$$

commutes, so that ${}^{g^{-1}}v = u$. Thus $v = {}^{g}u$, verifying our claim.

Next, if (X',τ') is an object of trans X and $\pi'\colon (Y,t) \to (FX',\delta\tau')$ is an arrow in trans Y, the triangle

$$\begin{array}{ccc}
 & Y & \\
{\scriptstyle\pi}\swarrow & & \searrow{\scriptstyle\pi'} \\
FX & \xrightarrow[\;F\xi\;]{} & FX'
\end{array}$$

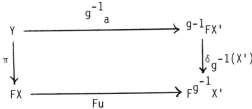

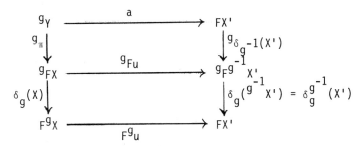

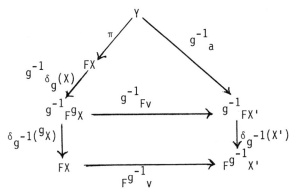

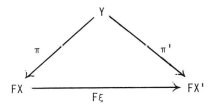

commutes for a unique arrow $\xi: X \to X'$ in X. It suffices to show that ξ is a transverse arrow $(X,\tau) \to (X',\tau')$. But, since π is a transverse arrow

$$F(\xi \circ \tau_g) \circ \delta_g(X) \circ {}^g\pi = F\xi \circ F\tau_g \circ \delta_g(X) \circ {}^g\pi = F\xi \circ (\delta\tau)_g \circ {}^g\pi$$

$$= F\xi \circ \pi \circ t_g = \pi' \circ t_g$$

and, since π' is a transverse arrow

$$F(\tau'_g \circ {}^g\xi) \circ \delta_g(X) \circ {}^g\pi = F\tau'_g \circ F{}^g\xi \circ \delta_g(X) \circ {}^g\pi$$

$$= F\tau'_g \circ \delta_g(X') \circ {}^gF\xi \circ {}^g\pi = (\delta\tau')_g \circ {}^g(F\xi \circ \pi)$$

$$= (\delta\tau')_g \circ {}^g\pi' = \pi' \circ t_g$$

by naturality of δ_g. Therefore $\xi\tau_g = \tau'_g{}^g\xi$; that is, ξ is a transverse arrow. □

Next, let A, B and C be G-categories and $(N,\nu): B \to C$ be a transversaled functor. Now the induced functor $N_*: B^A \to C^A$ is a G^{op}-functor, the induced natural isomorphisms $\nu_{g*}: {}^gN_* \to N_*^g$ are G^{op}-natural and N_*, seen as functor $\dot{B}^A \to \dot{C}^A$, is transversaled by $\dot{\nu}_*$. Thus trans N_* is defined and, in fact, is just the evident functor gotten by composing 1-cells. In particular one has

COROLLARY 6.2. Given 1-cells $(M,\mu): A \to B$ and $(T,\tau): A \to C$ and a 2-cell $\lambda: (T,\tau) \to (NM,\nu\mu)$ such that (M,λ) is initial in $(T{\downarrow}N_*)$, $((M,\mu),\lambda)$ is initial in $((T,\tau){\downarrow}\text{trans } N_*)$. □

Consider a G-category C, a small G-category J and a transversaled diagram Z in C over J. Noting that the diagonal $\Delta: C \to C^J$ is a $G \times G^{op}$-functor, we may define a *transverse limit* and a *trivial-transverse limit of* Z to be, respectively, representations of

$$[\text{trans } C^J](\text{trans } \Delta-,Z): (\text{trans } C)^{op} \to \text{Set}$$

$$[\text{trans } C^J](\text{triv-trans } \Delta-,Z): (\text{triv-trans } C)^{op} \to \text{Set}\,.$$

For instance, a trivial-transverse limit of a trivially transversaled diagram is a G-limit (and vice versa). We put

$$\text{trans-cone}(-,Z) = [\text{trans } C^J](\text{trans } \Delta-,Z)$$

$$\text{triv-trans-cone}(-,Z) = [\text{trans } C^J](\text{triv-trans } \Delta-,Z)$$

and refer to *transverse cones* and *trivial-tranverse cones*, respectively.

Next we study transversaled functors $C \to \mathrm{par}(G\text{-Set})$. The supereminent example, naturally, is $C(C,-)$, where C is transversaled by ϕ, say -- for then, as is readily checked, $C(C,-)$ is transversaled by the family $C(\phi_g^{-1}, {}^g-)$. Another example is thus the composite

$$\mathrm{cone}(-,Z)\colon C^{op} \xrightarrow{\Delta^{op}} (\dot{C}^J)^{op} \xrightarrow{\dot{C}^J(-,\dot{Z})} \mathrm{par}(G\text{-Set})$$

where $\dot{Z}$ is the transversaled diagram gotten from Z via the canonical isomorphism $\mathrm{trans}\ C^J \to \mathrm{trans}\ \dot{C}^J$.

Now let $(T,\tau)\colon C \to \mathrm{par}(G\text{-Set})$ be an arbitrary transversaled functor. We consider trans T as a G-Set-valued functor via the canonical equivalence of Proposition 4.2; that is, we identify $\mathrm{trans}T(C,\phi)$ with the G-set TC under the $\circ$-operation induced by the system of isomorphisms $\tau\phi$ at TC to wit:

$$g \circ x = (\tau\phi)_g {}^g x, \qquad x \in TC,\ g \in G.$$

This enables us, analogously as in Section 4, to define a subfunctor ${}^G T$: trans $C \to$ Set of trans T by taking ${}^G T(C,\phi)$ to be the fixed point set of $\mathrm{trans}T(C,\phi)$:

$${}^G T(C,\phi) = \{x \in TC \mid (\tau\phi)_g {}^g x = x,\ \text{all } g \in G\}.$$

It is computationally useful to observe that ${}^G T(C,\phi)$ may equally be described as the set of all x for which ϕ is a transversal for the object (C,x) of the G-category $(\dagger{\downarrow}T)$, $\dagger$ being the one-point trivial G-set. Indeed, where $*$ is the one-point set:

PROPOSITION 6.3. The comma categories $(*{\downarrow}{}^G T)$ and $(\dagger{\downarrow}\mathrm{trans}\ T)$ are identical for any par(G-Set)-valued transversaled functor T. □

The restriction of ${}^G T$ to triv-trans C is denoted triv-${}^G T$. For instance, when $T = C(C,-)$, C transversaled by ϕ, ${}^G T = [\mathrm{trans}\ C]((C,\phi),-)$; and then, for ϕ trivial, triv-${}^G T = [\text{triv-trans}\ C](C,-)$. Also, when $T = \mathrm{cone}(-,Z)$, ${}^G T = \text{trans-cone}(-,Z)$ and triv-${}^G T = \text{triv-trans-cone}(-,Z)$. Finally, it is important to note that ${}^G T$ and triv-${}^G T$ are functorial in T, meaning

$${}^G-\colon \mathrm{trans}(\mathrm{par}(G\text{-Set}))^C \to \mathrm{Set}^{\mathrm{trans}\ C}$$

$$\text{triv-}{}^G-\colon \mathrm{trans}(\mathrm{par}(G\text{-Set}))^C \to \mathrm{Set}^{\text{triv-trans}\ C}$$

are functors; for, if $\omega\colon T \to S$ is a transverse arrow in $(\mathrm{par}(G\text{-Set}))^C$,

${}^{G}\omega\colon {}^{G}T \to {}^{G}S$ defined by commutativity of

$$\begin{array}{ccc} {}^{G}T(C,\phi) & \xrightarrow{\;{}^{G}\omega(C,\phi)\;} & {}^{G}S(C,\phi) \\ \cap & & \cap \\ TC & \xrightarrow[\;\omega_C\;]{} & SC \end{array}$$

is an arrow in $\mathrm{Set}^{\mathrm{trans}\ \mathcal{C}}$ and compositions and identities are preserved. Concerning these functors, see Theorem 6.9.

By a *transversaled representation of* T we mean a transversaled object C of $\mathcal{C}$ together with a transverse natural isomorphism $\mathcal{C}(C,-) \to T$. When C is trivially transversaled we use the terminology *trivially transversaled representation*. Note that, whereas for trivially transversaled T (= G-functors) the former notion was heretofore undefined, the latter notion is exactly that of stable representations. *Transversaled limits* and *trivially transversaled limits* of Z (Z as above) are defined to be, respectively, transversaled representations and trivially transversaled representations of cone$(-,Z)$.

We should translate into more usual terms: a transversaled limit of Z is a transverse cone(= 2- cell) $C \to Z$ such that every ordinary cone $D \to Z$ has a unique (ordinary) factorization through $C \to Z$. A trivially transversaled limit of Z is a trivial-transverse cone(= 2 -cell with trivially transversaled domain) to Z through which every cone to Z factors uniquely. Thereby, a transverse limit of Z is a transverse cone to Z through which every transverse cone to Z has a unique transverse factorization, and a trivial-transverse limit of Z is a trivial-transverse cone to Z through which every trivial-transverse cone to Z has a unique transverse factorization.

The translation for transversaled limits is an immediate consequence of the fact that Yoneda for G-functors -- Lemma 4.4 -- remains valid for transversaled functors; that is, 2- cells $\mathcal{C}(C,-) \to T$ correspond bijectively, under the Yoneda map, to elements of $({}^{G}T)C$ (the proof being only slightly less routine than that of 4.4). In particular, by Corollary 2.5(i), transversaled representations of T amount to initial objects of the G-category $(\dagger\downarrow T)$ -- of course, trivially transversaled representations are essentially stable (= trivially transversaled) initial objects.

A fortiori, any transversaled representation of T can be regarded as a representation of the corestriction of T to Set. Conversely, we have

PROPOSITION 6.4. Let T be a transversaled par(G-Set)-valued functor on C and let η: $C(C,-) \to T$ be a representation of T viewed as Set-valued functor. Then there is a unique transversal for C such that η is a transversaled representation of T.

Proof. Given "transversaled Yoneda" above, this is just 2.5 again. □

This result is readily applicable to limits of transversaled diagrams over trivial G-categories. For example, suppose (X,ζ) is a transversaled object of a G-category X and that e is a transverse idempotent endomorphism of (X,ζ). Let

$$X \xrightarrow{q} Y \xrightarrow{p} X, \qquad pq = e,\ qp = 1$$

be a splitting of e (i.e., p is equalizer of e and 1_X). By Proposition 6.4, this is a transversaled splitting of e for a unique transversal t for Y; namely, the t making $p:(Y,t) \to (X,\zeta)$ a transverse arrow, in which case $q:(X,\zeta) \to (Y,t)$ is a transverse arrow too. By the next result -- a generalization of Theorem 2.7 -- if X is stably closed, e has a trivially transversaled splitting; that is, Y, p and q can be chosen so that

$$\begin{array}{ccccc} gX & \xrightarrow{gq} & gY & \xrightarrow{gp} & gX \\ {\scriptstyle \zeta_g}\downarrow & & \| & & \downarrow{\scriptstyle \zeta_g} \\ X & \xrightarrow[q]{} & Y & \xrightarrow[p]{} & X \end{array}$$

commutes for all g.

PROPOSITION 6.5. The following are equivalent properties of a G-category C.

(i) Every transversally representable transversaled functor $C \to$ par(G-Set) is trivially transversally representable.

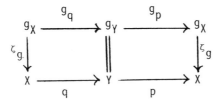

(ii) $C(C,-)$ is trivially transversally representable whenever C is transversaled .

(iii) C is stably closed.

Proof. (i)$\Rightarrow$(iii). Immediate by 6.4.

(iii)$\Rightarrow$(ii). Suppose C is transversaled by ϕ. Since, by 2.2, ϕ has a limit, ϕ has a stable limit. However, a stable limit of ϕ is in effect a transverse isomorphism of the form $f\colon (C,\phi) \to (D,1)$, by 2.2. So $f^*\colon C(D,-) \to C(C,-)$ is a trivially transversaled representation of $C(C,-)$.

(ii)$\Rightarrow$(i). If $C(C,-) \to T$ is a transversaled representation of T and $C(D,-) \to C(C,-)$ is a trivially transversaled representation of $C(C,-)$, then the composite $C(D,-) \to C(C,-) \to T$ is a trivially transversaled representation of T. □

Thus:

COROLLARY 6.6. C is stably closed if and only if every transversaled diagram in C which has a limit has a trivially transversaled limit. □

In Section 2 we promised to explain what it meant for every system of isomorphisms in a G-category C to have a G-limit. We saw there that if ϕ is a system of isomorphisms at C, regarded as G-diagram in C over G, a cone $\phi \to D$ is effectively an arrow $C \to D$. In the same way a transverse cone $\kappa\colon \phi \to (D,\Psi)$ is effectively a transverse arrow $f\colon (C,\phi) \to (D,\Psi)$, for since f must be the 1th component of κ

$$\Psi_g {}^g\kappa_h = \Psi_g {}^g(f\phi_h) = \Psi_g {}^g f {}^g\phi_h = f\phi_g {}^g\phi_h = f\phi_{gh} = \kappa_{gh}$$

which is equivalent to commutativity of

$$\begin{array}{ccc} {}^g\phi & \xrightarrow{\ {}^g\kappa\ } & {}^gD \\ \Vert & & \downarrow{\scriptstyle \Psi_g} \\ \phi^g & \xrightarrow[\ \kappa^g\ ]{} & D \end{array} .$$

In particular we see that the identity transverse arrow $(C,\phi) \to (C,\phi)$ is a transverse colimit of ϕ (i.e., ϕ_g is g^{th} component of a colimiting transverse cone). This proves part of

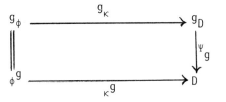

THEOREM 6.7. Every system of isomorphisms in a G-category X has a G-limit if and only if every G-diagram in X which has a transverse limit has a G-limit.

The other part is a consequence of the following counterpart of Proposition 6.5 concerning representable functors on trans C.

PROPOSITION 6.8. The following statements are equivalent.

(i) The restriction of every representable functor trans $C \to$ Set to triv-trans C is representable.

(ii) For all $(C,\phi) \in$ trans C, the restriction of [trans C]$((C,\phi),-)$ to triv-trans C is representable.

(iii) Every transversaled diagram in C having a transverse colimit has a trivial-transverse colimit.

(iv) Every system of isomorphisms in C has a G-colimit.

Proof. The equivalence of (i) and (ii) is obvious, (ii) is equivalent to (iv) by Yoneda, (iii) is a special case of (i), and we just saw that (iii) implies (iv). □

Strictly speaking there was no need for the explicit computation above of a transverse colimit for a system of isomorphisms: By Lemma 2.2 systems of isomorphisms have limits and any transversaled diagram which has a limit, has a transversaled limit and, thus, has a transverse limit, by Proposition 6.1. In fact, combining Propositions 6.1 and 6.3 one gets that ${}^{G}T$ is representable for any transversaled functor T. A better proof is afforded by the next partial generalization of Theorem 4.7.

THEOREM 6.9. Let T be a transversaled functor $C \to$ par(G-Set) and let C be an object of C. Then the triangles in Figure 1 are commutative diagrams of natural isomorphisms, all unlabled arrows being Yoneda maps. In particular, if C is transversaled by ϕ and η is a transversaled representation of T by C then ${}^{G}\eta$ is a representation of ${}^{G}T$ by (C,ϕ); and if C is trivially transversaled and η a trivially transversaled representation of T by C, triv-${}^{G}\eta$ is a representation of triv-${}^{G}T$ by C. □

Figure 1

C tranversaled by ϕ:

$$\begin{array}{ccc}
[\mathrm{trans}(\mathrm{par}(G\text{-}\mathrm{Set}))^{\mathcal{C}}](\mathcal{C}(C,-),T) & \xrightarrow{\;G_{-}\;} & \mathrm{Set}^{\mathrm{trans}\ \mathcal{C}}([\mathrm{trans}\ \mathcal{C}]((C,\phi),-),{}^{G}T) \\
& \searrow \qquad \swarrow & \\
& ({}^{G}T)(C,\phi) &
\end{array}$$

C trivially transversaled:

$$\begin{array}{ccc}
[\mathrm{trans}(\mathrm{par}(G\text{-}\mathrm{Set}))^{\mathcal{C}}](\mathcal{C}(C,-),T) & \xrightarrow{\;\mathrm{triv}\text{-}G_{-}\;} & \mathrm{Set}^{\mathrm{triv\text{-}trans}\ \mathcal{C}}([\mathrm{triv\text{-}trans}\ \mathcal{C}]((C,1),-),\mathrm{triv}\text{-}{}^{G}T) \\
& \searrow \qquad \swarrow & \\
& (\mathrm{triv}\text{-}{}^{G}T)(C,1) &
\end{array}$$

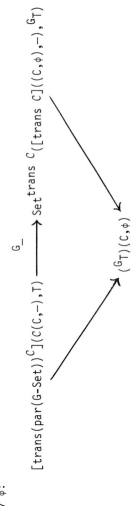

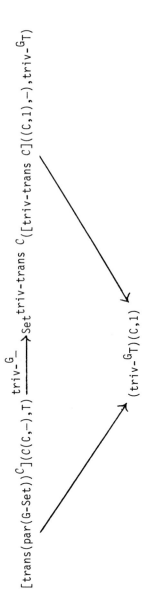

For example, in a complete G-category transverse limits and transversaled limits are the same and, in a complete stably closed G-category (such as par(G-Set)), trivial-transverse limits and trivially transversaled limits are the same. Especially, if Σ is a transversaled diagram in par(G-Set), mimicking the argument of Proposition 4.9, we find that trans-cone($G\cdot *,\Sigma$) has a unique G-set structure such that the cone σ: trans-cone($G\cdot *,\Sigma$) $\to \Sigma$ defined by $\sigma_j x = x_j{}^*$ for $j \in \mathrm{dom}\Sigma$ is a (trivially) transversaled limit of Σ.

Just as partial G-hom-functors need not preserve G-limits, transversaled partial hom-functors need not preserve transverse limits. To rectify this we introduce the concept of derived-transverse limits of transversaled diagrams Σ in par(G-Set) over J: Denote the composite

$$\mathrm{triv(G\text{-}Set)} \xrightarrow{\subset} \mathrm{G\text{-}Set} \xrightarrow{\Delta} \mathrm{trans\ par(G\text{-}Set)} \xrightarrow{\mathrm{trans}\ \Delta} \mathrm{trans(par(G\text{-}Set))}^J$$

by Δ' (see 4.2), where triv-(G-Set) connotes the full subcategory of G-Set consisting of the trivial G-Sets. Accordingly we denote the contravariant functor

$$[\mathrm{trans(par(G\text{-}Set))}^J](\Delta'-,\Sigma)\colon (\mathrm{triv(G\text{-}Set)})^{op} \longrightarrow \mathrm{Set}$$

by trans-cone'$(-,\Sigma)$ -- we speak of *derived-transverse cones* and *derived-transverse limits*, the latter being representations of trans-cone'. Now if X'_ϕ is the fixed point set of the G-set X_ϕ corresponding to a system of isomorphisms ϕ at a partial G-set (X,W), since for a transverse cone $((X,W),\phi) \to \Sigma$ the composite

$$X'_\phi \subset X_\phi \xrightarrow{\cong} ((X,W),\phi) \longrightarrow \Sigma$$

is a derived-transverse cone, we get a natural map

$$-'\colon \text{trans-cone}(((X,W),\phi),\Sigma) \to \text{trans-cone}'(X'_\phi,\Sigma)$$

from transverse cones to derived-transverse cones. Of course, $-'$ takes transverse limits to derived-transverse limits.

LEMMA 6.10. Suppose that Z is a transversaled diagram in C and κ: $L \to Z$ is a transverse cone. Then κ is a transverse limit if and only if the restriction $C(C,\kappa)|$: trans$C(C,L) \to C(C,Z-)$ is a derived-transverse limit for all transversaled objects C of C.

Naturally, the analogous result for trivial-transverse cones κ is valid too.

To prove Lemma 6.10, consider the commutative diagram

$$\begin{array}{ccccc}
\text{trans}\mathcal{C}(C,L) & & \xrightarrow{\quad ({}^{G}\kappa_*)_C \quad} & & \text{trans-cone}\,(C,Z) \\
& \overset{\subset}{\searrow} & & \overset{\supset}{\swarrow} & \\
\mathcal{C}(C,\kappa)|\ \searrow & \mathcal{C}(C,L) & \xrightarrow{\quad \kappa_{*C} \quad} & \text{cone}(C,Z) & \swarrow\ \delta_C| \\
& \downarrow{\scriptstyle \mathcal{C}(C,\kappa)} & & \downarrow{\scriptstyle \delta_C} & \\
& \mathcal{C}(C,Z-) & = & \mathcal{C}(C,Z-) &
\end{array}$$

of transverse natural transformations of transversaled functors of $J = \text{dom}Z$, natural in $C \in$ trans $\mathcal{C}$, where $\kappa_*:\mathcal{C}(-,L) \to \text{cone}(-,Z)$ is the transverse natural transformation induced via Theorem 6.9 by κ and δ_C is defined by $(\delta_C)_j\gamma = \gamma_j$, $j \in J$. But, as argued in Section 4, δ_C may be written as

$$\text{cone}(C,Z) \xrightarrow{f_C} \text{trans-cone}(G\cdot{*},\mathcal{C}(C,Z-)) \xrightarrow{\sigma_C} \mathcal{C}(C,Z-)$$

where f_C is bijective--except here the definition of f_C is more intricate:

$$(f_C\gamma)_j{}^{g*} = \zeta_g({}^{g^{-1}}j) \circ {}^{g}\gamma_{g_j^{-1}} \circ \phi_g^{-1}, \qquad Z = (Z,\zeta), \quad C = (C,\phi).$$

We get, nevertheless, that δ_C is a transverse limit and thus, inasmuch as $\delta_C| = \delta'_C$, $\delta_C|$ is a derived-transverse limit. So, for all transversaled C, $\mathcal{C}(C,\kappa)|$ is a derived-transverse limit if and only if $({}^{G}\kappa_*)_C$ is an isomorphism; that is, according to Theorem 6.9, if and only if κ is a transverse limit.

COROLLARY 6.11. If T is a transversally representable transversaled functor and $\Theta:L \to Z$ is a transverse limit in the domain of T, then $(T\Theta)|:({}^{G}T)L \to TZ$ is a derived-transverse limit. The same is true if T is trivially transversally representable and Θ is a trivial-transverse limit.

Proof. The proof consists, for the most part, in observing that

$$\begin{array}{ccccc}
[\text{trans}\ \mathcal{C}](C,L) & & \xrightarrow{\quad ({}^{G}\eta)_L \quad} & & ({}^{G}T)L \\
& \overset{\subset}{\searrow} & & \overset{\supset}{\swarrow} & \\
\mathcal{C}(C,\Theta)|\ \searrow & \mathcal{C}(C,L) & \xrightarrow{\quad \eta_L \quad} & TL & \swarrow\ (T\Theta)| \\
& \downarrow{\scriptstyle \mathcal{C}(C,\Theta)} & & \downarrow{\scriptstyle T\Theta} & \\
& \mathcal{C}(C,Z-) & \xrightarrow{\quad \eta Z \quad} & TZ &
\end{array}$$

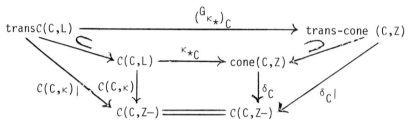

is a commutative diagram, where η is a transversaled representation of T by C. □

THEOREM 6.12. A transversaled functor preserves transverse limits if it has a left adjoint and preserves trivial-transverse limits if it has a left trivially transversaled adjoint.

This is an easy consequence of the two results that precede it, Proposition 5.2, and

THEOREM 6.13. Let N: $\mathcal{Q} \to \mathcal{P}$ be a transversaled functor.

(i) N has a left transversaled adjoint if and only if, for every subgroup H of G and every H-transversaled object P of $\mathcal{P}$, $\mathcal{P}(P,N-)$ is H-transversally representable.

(ii) N has a left trivially transversaled adjoint if and only if $\mathcal{P}(P,N-)$ is G(P)-trivially transversally representable for all $P \varepsilon \mathcal{P}$.

Proof. The second part follows by 5.5 and the Yoneda lemma for transversaled functors, while the first part is an immediate consequence of 5.2 and 6.4. □

Notice that, since $\mathcal{P}$ can have non-transversaled objects, $\mathcal{P}(P,N-)$ being transversally representable for every transversaled P in $\mathcal{P}$ does not ensure N having a left adjoint.

7. REFLECTIONS AND STABLE REFLECTIONS

Let C and J be G-categories where J is small. There is another way of looking at G-limits and stable limits that we have ignored up till now. For this, consider the commutative square

$$\begin{array}{ccc} C & \xrightarrow{\quad \Delta \quad} & \dot{C}^J \\ \cup & & \cup \\ \text{stab } C & \xrightarrow[\underline{\Delta}]{\quad\quad} & \text{stab } \dot{C}^J \end{array} \tag{1}$$

where Δ and $\underline{\Delta}$ are the respective diagonal and stable diagonal functors. In that we wish to deal with reflective subcategories rather than coreflective ones, we study colimits instead of limits. In particular, we notice that every G-diagram in C over J has a G-colimit if and only if $\underline{\Delta}$ has a left adjoint.

When J is the discrete G-category G, we view objects or arrows of $\dot{C}^J$ as families $\{*_g\}_{g\varepsilon G}$, where the $*_g$ are, respectively, objects or arrows of C. It is important to note that the family $\{*_g\}$ is stable exactly when $*_g = {}^g*_1$ for all g.

These observations lead to

THEOREM 7.1. The stable subcategory of C is reflective precisely when C has G-indexed G-coproducts. Moreover if C is stably closed, the following statements are equivalent, where E denotes the restriction of evaluation at the identity of G stab $C^G \to C$.

(i) Stab C is reflective;

(ii) insertion stab $C \to C$ is tripleable;

(iii) E has a left adjoint;

(iv) E is monadic.

Proof. We have the commutative triangle:

$$\begin{array}{ccc} \text{stab } C & \xrightarrow{\underline{\Delta}} & \text{stab } C^G \\ & \cap & \swarrow \\ & C & \end{array} \qquad (2)$$

the unlabeled arrow being restriction of evaluation at 1. But this restriction is clearly an isomorphism, proving the initial assertion. For the rest, note that the triangle (2) remains commutative when G is replaced by G (the G-groupoid of § 2). Thereto, assuming C to be stably closed, $\underline{\Delta}$: stab $C \to$ stab C^G is an equivalence, by 2.8. But, by 2.1 and Beck's theorem (i.e., [22, Theorem 1, p. 147]), E is monadic if and only if it has a left adjoint; and compositions of monadic functors with equivalences are tripleable (see 8.9(ii)). □

We mention that the implication (i)$\Rightarrow$(ii) is also an easy consequence of PTT (= precise tripleability theorem -- cf. 8.8) using Proposition 1.5(iii) and either of Corollaries 1.6 or 1.3.

For a different perspective on Theorem 7.1 see the discussion below Theorem 9.1. Also, concerning the necessity of the condition in the next result, see Theorem 9.5.

PROPOSITION 7.2. A sufficient condition for stab C to be reflective is that restriction of evaluation at 1 stab $C^G \to C$ has a left adjoint which factors through the stable diagonal stab $C \to$ stab C^G.

Proof. Let E and $\underline{\Delta}$ be the respective restriction and diagonal, and let L be a left adjoint of E such that L = $\underline{\Delta}$M for some M: $C \to$ stab C. Then, since $\underline{\Delta}$ is fully faithful

$$[\text{stab } C](MC,D) \cong [\text{stab } C^G](\underline{\Delta}MC,\underline{\Delta}D)$$

$$= [\text{stab}C^G](LC,\underline{\Delta}D) \cong C(C,E\underline{\Delta}D) = C(C,D)$$

naturally for $C \in C$ and D $\in$ stab C. □

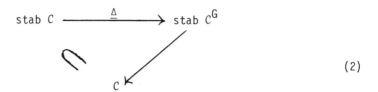

Theorem 7.1 has a counterpart concerning the evident forgetful functor trans $C \to C$; namely, trans $C \to C$ has a left adjoint if and only if C has G-indexed transverse coproducts, in which case trans $C \to C$ is monadic. In fact, in the commutative triangle

$$\begin{array}{ccc} \text{trans } C & \xrightarrow{\text{trans } \Delta} & \text{trans } C^G \\ & \searrow \quad \swarrow & \\ & C & \end{array} \tag{3}$$

with trans $C^G \to C$ the composite

$$\text{trans } C^G \longrightarrow C^G \xrightarrow{E_1} C \qquad (E_1 = \text{evaluation at } 1)$$

we claim that trans $C^G \to C$ is an equivalence. But trans $C^G \to C$ plainly has as right inverse the composite $C \to$ triv-trans $C^G \subset$ trans C^G, where the arrow is the inverse of the restriction of the evaluation E_1 to triv-trans C^G. Now an object of trans C^G amounts to a pair of G-indexed families, say $\{C_g\}$ and $\{t_g\}$, where the C_g are objects of C and the t_g are isomorphisms ${}^gC_1 \to C_g$ in C with $t_1 = 1$ (the t_g define a transversal τ via $\tau_g(h) = t_{gh} \circ {}^gt_h^{-1}$). In particular

$$\begin{array}{ccc} {}^gC_1 & \xrightarrow{{}^gt_1^{-1}} & {}^gC_1 \\ {\scriptstyle t_g}\big\downarrow & & \big\| \\ C_g & \xrightarrow[t_g^{-1}]{} & {}^gC_1 \end{array}$$

is a commutative square. So triv-trans C^G is an equivalent subcategory of trans C^G (cf. 7.4), thus verifying our claim.

Incidentally, for the record, we see that a G-indexed transverse coproduct in C amounts to a transversaled object (D,Ψ) of C and an arrow $\Theta\colon C \to D$ in C such that for any transversaled object (D',Ψ') and arrow $\Theta'\colon C \to D'$ there is a unique transverse arrow $u\colon (D,\Psi) \to (D',\Psi')$ such that $u\Theta = \Theta'$. Thereby, we may compute that C has G-indexed trivial-transverse coproducts if and only if C has G-indexed G-coproducts (i.e., iff stab C is reflective). Alternatively, we may note that (3) still commutes when trans C is replaced by triv-trans C

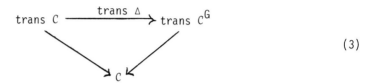

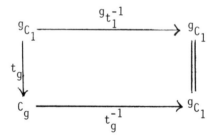

and trans Δ by triv-trans Δ -- so that triv-trans $C \to C$ has a left adjoint exactly when C has G-indexed trivial-transverse coproducts -- and then we may use that triv-trans C^G is an equivalent subcategory of trans C^G.

Many of the results of this section and those that follow are, as Theorem 7.1, special cases of results valid in Trans-Cat. Usually, however, we stick to G-Cat since results there are easier to state, prove and understand and, we think, of greater precedence.

When C has stable G-indexed coproducts -- automatically implying (e.g., 1.2 and 7.1) stab C is a reflective subcategory of C -- we say that stab C is a *stably reflective* subcategory of C. This means that for each $C \varepsilon C$ there is an object SC of stab C and an arrow μ_C: $C \to SC$ such that the family $\{{}^g\mu_C: {}^gC \to SC\}_{g\varepsilon G}$ is a coproduct; and then S is object function of a unique functor S: $C \to$ stab C left adjoint to insertion with unit having C^{th} component μ_C. Such adjunctions we term *stable reflections*. Note that an easily recognizable way for there to be stable reflections is for C to be stably closed and to have G^{th} *copowers*; that is, coproducts of the form $\coprod_{g\varepsilon G} {}^gC$ exist for any $C \varepsilon C$. Assuming stab C is stably reflective, by a *stable reflector* we mean a left adjoint for inclusion of stab C in C. Thus a stable reflector is in essence any stable endofunctor on C stably naturally isomorphic to the stable endofunctor whose corestriction to stab C is the above S.

We should mention that stable reflection typically cannot be a G-adjunction. An elementary way of seeing this is to note that if a stable reflector, R say, is a G-functor, since R preserves G-indexed coproducts, if $G \neq 1$ it follows there is at most one arrow from RC to a stable object for any object C. This, incidentally, renders easy the construction of examples of the type suggested at the end of Section 5.

Vis a vis the second part of Theorem 7.1, rephrasing Proposition 2.10:

PROPOSITION 7.3. If C has stable reflections, stab C has coequalizers, and these remain coequalizers in C (i.e., stab C coreflective), then C is stably closed. □

In truth this result is better thought of as giving sufficient conditions for insertion of stably reflective stable subcategories of G-categories to be tripleable: by Theorem 9.7, such tripleable insertions exist only for stably closed G-categories.

Next, referring to (1), one sees by Theorem 4.6 that Δ has a left G-adjoint if and only if each diagram Z in C over J has a G(Z)-stable colimit. This entails, by Corollary 4.8, that $\underline{\Delta}$ has a left adjoint, so that im $\underline{\Delta}$ is some sort of special reflective subcategory of stab C^J (each component of a unit of reflection is a stable J-indexed colimit). More to the point, when J = G, stab C is a stably reflective subcategory of C. Now, we know of no intrinsic conditions on C equivalent to $\Delta\colon C \to \dot{C}^G$ having a left G-adjoint; however, we do know exactly when Δ is "G-tripleable" in terms of properties of C. We leave this to the next section which deals systematically with G-tripleability. The result we have in mind, though (see 8.6), is based on

THEOREM 7.4. If G operates without fixed points on the objects of J then $\dot{C}^J$ is hereditarily stably closed.

Proof. It suffices to show that $\dot{C}^J$ is stably closed or, what is the same thing, that all systems of isomorphisms $(Z,\phi) \in \mathrm{stab}(\dot{C}^J)^G$ are stabilizing. Thus, given (Z,ϕ), one must find a G-functor $V\colon J \to C$ and natural isomorphism $\omega\colon V \to Z$ such that

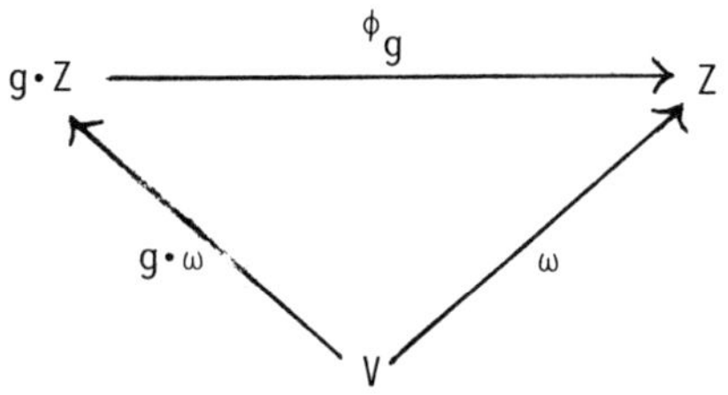

(4)

commutes. For this, choose a representative object in each G-orbit of objects of J and, for $j \in J$, denote by $j\dagger$ the chosen representative of the orbit of j. Then, inasmuch as G acts without fixed points, $j = {}^x j^\dagger$ for a unique $x \in G$. Put $V_j = (x\cdot Z)_j$ and, if $a\colon j \to k$, $k = {}^y k^\dagger$, put

$$Va = (y \cdot \phi_{y^{-1}x})_k \circ (x \cdot Z)a\colon V_j \to V_k.$$

Now if $b\colon k \to l$, $l = {}^{z}1^{\dagger}$, since

$$(z \cdot \phi_{z^{-1}y}) \cdot (y \cdot \phi_{y^{-1}x}) = z \cdot (\phi_{z^{-1}y} \cdot (z^{-1}y \cdot \phi_{y^{-1}x})) = z \cdot \phi_{z^{-1}x},$$

$$Vb \circ Va = (z \cdot \phi_{z^{-1}y})_l \circ (y \cdot Z)b \circ (y \cdot \phi_{y^{-1}x})_k \circ (x \cdot Z)a$$

$$= (z \cdot \phi_{z^{-1}y})_l \circ (y \cdot \phi_{y^{-1}x})_l \circ (x \cdot Z)b \circ (x \cdot Z)a$$

$$= (z \cdot \phi_{z^{-1}x})_l \circ (x \cdot Z)ba = Vba$$

by naturality of $y \cdot \phi_{y^{-1}x}$. Thus since, clearly, $V1 = 1$, V is a functor. Furthermore

$${}^{g}V_j = {}^{g}(x \cdot Z)_j = (gx \cdot Z)_{{}^{g}j} = V_{{}^{g}j};$$

$${}^{g}(Va) = {}^{g}(y \cdot \phi_{y^{-1}x})_k \circ {}^{g}(x \cdot Z)a$$

$$= (gy \cdot \phi_{y^{-1}x})_{{}^{g}k} \circ (gx \cdot Z)^{g}a$$

$$= (gy \cdot \phi_{(gy)^{-1}gx})_{{}^{g}k} \circ (gx \cdot Z)^{g}a = V({}^{g}a).$$

So V is a G-functor.

To finish, define ω by $\omega_j = (\phi_x)_j$ where, as before, x is defined by ${}^{x}j^{\dagger} = j$. Naturality of ω is a routine consequence of naturality of the ϕ_g and their multiplication rule. Commutativity of (4) is just the multiplication rule:

$$(\phi_g)_j \circ (g \cdot \omega)_{\dot{j}} = (\phi_g)_j \circ (g \cdot \phi_{\dot{g}^{-1}x})_j = (\phi_x)_j = \omega_j. \quad \square$$

COROLLARY 7.5. If C^J has G^{th} copowers then C^J has stable reflections, provided G acts without fixed points on $\text{obj}J$. $\square$

Thus $\dot{C}^G$ and $\dot{C}^G$ have stable reflections whenever C has G-indexed coproducts. Notice that, when the action of G on C is trivial, stab $\dot{C}^G$ is the image of the diagonal $C \to \dot{C}^G$. Consequently, for a trivial G-category C, the stable subcategory of $\dot{C}^G$ is reflective when and only when C has G-indexed coproducts so that, in particular, C has G-indexed coproducts if and only if $\dot{C}^G$ has G^{th} copowers.

If C is any G-category with G-indexed coproducts, one may construct a stable reflector on $\dot{C}^G$ directly to wit: It is enough, given a G-indexed family of objects C_g of C, to construct an object D of C and a G-indexed family of arrows $\iota_g\colon C_g \to {}^gD$ such that the family $\{{}^h\iota_{h^{-1}g}\colon {}^hC_{h^{-1}g} \to {}^gD\}_{h\varepsilon G}$ is a coproduct for each fixed g. To do this, let $D = \amalg^k C_{k^{-1}}$ and let, say, u_k be the kth injection. Then, since $\{u_{g^{-1}h}\colon {}^{g^{-1}h}C_{h^{-1}g} \to D\}_{h\varepsilon G}$ is a coproduct for each g, $\{{}^gu_{g^{-1}h}\colon {}^hC_{h^{-1}g} \to {}^gD\}_{h\varepsilon G}$ is a coproduct for each g. But then, if one sets $\iota_g = {}^gu_{g^{-1}}$, $\iota_g \varepsilon C(C_g, {}^gD)$ and ${}^h\iota_{h^{-1}g} = {}^h({}^{h^{-1}g}u_{(h^{-1}g)^{-1}}) = {}^gu_{g^{-1}h}$.

Now, we know that the □-action of G on C^G is such that $E_1|\colon \text{stab } C^G \to C$ is a G-isomorphism. We infer:

THEOREM 7.6. Every G-category with G-indexed coproducts is a reflective G-subcategory of an hereditarily stably closed G-category with G-indexed coproducts (thus with stable reflections). Moreover, every category with G-indexed coproducts may be viewed as the stably reflective stable subcategory of some hereditarily stably closed G-category with G-indexed coproducts. □

Next, in the opposite extreme to the above, we consider the case when the action of G on J is trivial. Here C^J need not be stably closed. However, each evaluation $E_j\colon C^J \to C$ is a G-functor. It follows that, for any G-diagram V in C^J, $E_j(\lim_G V) = \lim_G(E_jV)$ whenever the pointwise G-limits exist. Thus we have

PROPOSITION 7.7. Suppose G acts trivially on J. Then C^J is as G-complete or as stably complete as C is, respectively, G-complete or stably complete. In particular, when C has stable reflections, C^J does too. □

COROLLARY 7.8. If J is a trivial G-category and C is stably closed then C^J is stably closed.

Proof. Use 2.7. □

We finish the section by studying the ramifications concerning the unit of stable reflection being split monic. Not the least reason this is important is, as shall be seen in Lemma 10.9, the condition implies stable reflectors have descent type. Initially we will regard C^C as a G-category with exponential C seen as trivial G-category (and base C in its given G-structure). Thereupon, assuming stab C to be stable reflective, a stable reflecttion for C can be described as a stable object R of C^C together with an arrow $\mu: 1_C \to R$ in C^C such that the G-indexed family of arrows ${}^g\mu: {}^g\_ \to R$ is a coproduct. Retractions of this μ correspond bijectively to G-indexed families of natural transformations $\delta_g: {}^g\_ \to 1_C$ with $\delta_1 = 1$. The simplest case occurs when $\{\delta_g\}$ is a "Kronecker family"; that is, C has zero arrows (e.g., C has a null object or is an Ab-category) and $\delta_g = 0$ whenever $g \neq 1$. In this case, because the corresponding retraction μ' satisfies

$$ {}^g\mu' \cdot {}^h\mu = {}^g(\mu' \cdot {}^{g^{-1}h}\mu) = {}^g\delta_{g^{-1}h} $$

μ' is characterized as the retraction of μ such that the ${}^g\mu'$ are the projections for the injections ${}^g\mu$ for all g. In this case we call μ' the *Kronecker retraction of* μ.

There are examples, for instance, par(G-Set) (cf. 7.11), where the δ_g form a system of isomorphisms. It is not difficult to see precisely when this happens:

PROPOSITION 7.9. (i) The following statements are equivalent, where E: stab $C^G \to C$ is restriction of evaluation at $1 \in G$.

(a) C^C has a system of isomorphisms at 1_C;

(b) the inclusion functor of some equivalent subcategory of C factors through E;

(c) E has a right inverse.

(ii) C^C has a stabilizing system of isomorphisms at 1_C if and only if stab C contains an equivalent subcategory of C.

Proof. (i). Plainly, (a) is equivalent to (c); and the implication (c)=> (b) is trivial. We prove that (b) implies (c): We are given a commutative triangle of the form

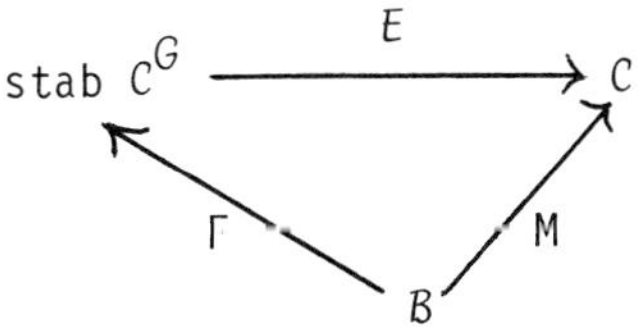

where M is an equivalence, and we are invited to find a functor $U: C \to$ stab C^G such that $EU = 1$. To this end, choose an inverse equivalence M' of M, set

$V = FM'\colon C \to \text{stab } C^G$, and let $\sigma\colon EV = EFM' = MM' \to 1$ be a natural isomorphism. On objects C, UC is taken to be the system of isomorphims at C given, for each $g \in G$, by $(UC)_g = \sigma_C (VC)_g {}^g\sigma_C^{-1}$. On arrows $\gamma\colon C \to D$, $U\gamma$ is taken to be the arrow $\gamma = \sigma_D \circ V\gamma \circ \sigma_C^{-1}\colon UC \to UD$ (naturality of σ). It is straightforward to check that U is as wished.

(ii). According to 2.4, the existence of a stabilizing system of isomorphisms at 1_C is equivalent to the existence of a stable endofunctor, K say, on C together with a natural isomorphism $K \to 1_C$. The result is a routine consequence of this. □

We should point out that, when C is stably closed, every system of isomorphisms at 1_C is stabilizing, by Corollary 7.8.

The foregoing is summarized in

COROLLARY 7.10. If C has zero arrows or possesses an equivalent subcategory whose insertion factors through restriction of evaluation at 1 $\text{stab } C^G \to C$, then units of stable reflection for C are split monic. □

For stably closed G-categories the latter case is unexpectedly ubiguitous:

THEOREM 7.11. Associated with each stably closed G-category X with stable reflections is a stably closed G-category Y with stable reflections and a monadic colimit creating G-functor $F\colon Y \to X$ such that a) stab Y contains an equivalent subcategory of Y, and b) F is comonadic if and only if stab X is coreflective (i.e., X has G-indexed G-products).

Proof. Let Y be the G-category $\text{stab } X^G$, so that restriction of evaluation at 1 $F\colon Y \to X$ is a G-functor. This F creates limits and colimits, by 2.1, and hence, by 1.7 and 2.7, Y is stably closed and admits stable reflections. Also we know, by 2.8, that the stable diagonal $\underline{\Delta}\colon \text{stab } X \to Y$ is an equivalence; and, as we saw below 2.8, stab Y contains im $\underline{\Delta}$. The rest follows by commutativity of

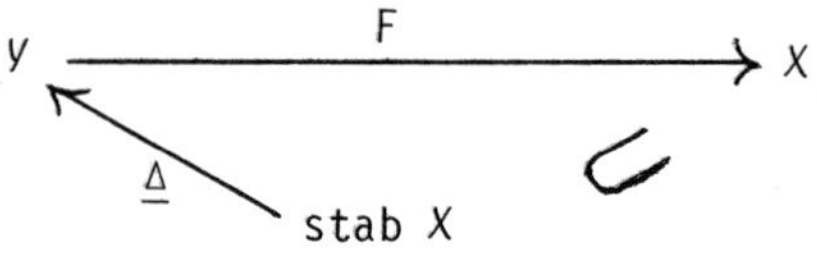

. □

It follows by a later result -- see Theorem 10.12 -- that if idempotents split in X, then stable reflectors on Y are cotripleable (this need not be true of stable reflectors on X as shown below 11.1).

Next, if R is a stable reflector on C and μ the unit of reflection, there is a unique family of stable isomorphisms $\rho_g: R^g \to R$ making

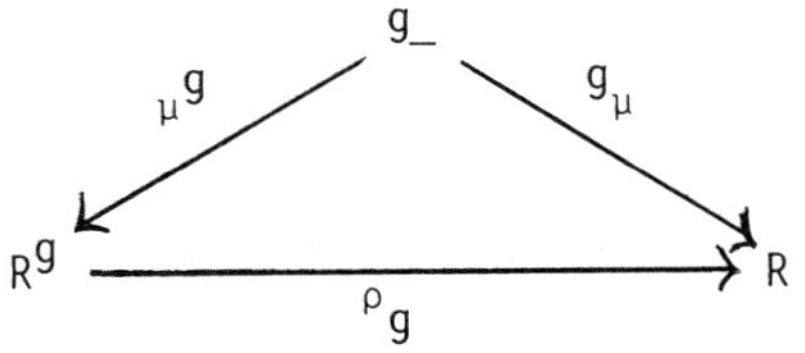

commutative. Otherwise put, the ρ_g constitute the cotransversal for R whose companion transversal is trivial (see 5.4). Given a retraction μ' of μ, in [14] the question arises as to whether μ' is compatible with the ρ_g in the sense that

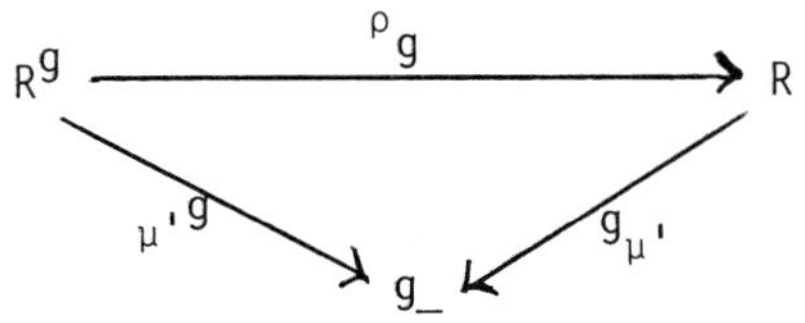

(5)

commutes. We mention this is the same, regarding R as cotransversaled by the family $\dot{\rho} = \rho_g^{g^{-1}}: R \to g\cdot R$, as asking whether $e = \mu \cdot \mu'$ is a cotransverse idempotent endomorphism of $(R,\{\dot{\rho}_g\})$, in which case $R \xrightarrow{\mu'} 1 \xrightarrow{\mu} R$ is a trivially cotransversaled splitting of e in $\dot{C}^C$ as discussed in Section 6.

LEMMA 7.12. (i) The diagram (5) commutes if and only if the δ_g defined by commutativity of

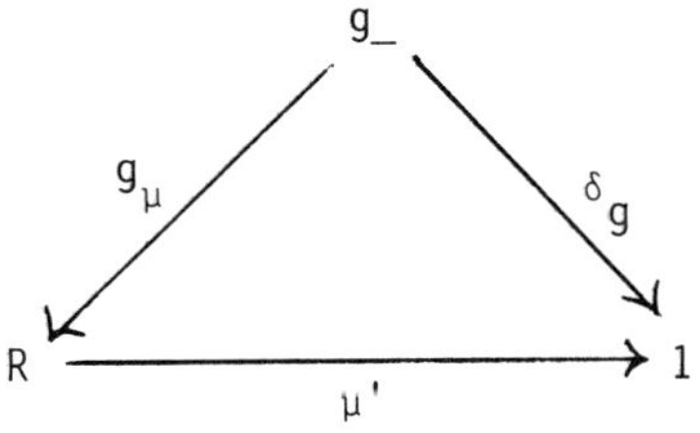

satisfy ${}^h\delta_{h^{-1}g} = \delta^h_{gh^{-1}}$.

(ii) If $\{\delta_g\}$ is either a Kronecker family or a system of isomorphisms at 1_C, then ${}^h\delta_{h^{-1}g} = \delta^h_{gh^{-1}}$.

Proof. (i). Inasmuch as the ${}^g\mu$ are injections for a coproduct, (5) commutes if and only if ${}^h\mu' \cdot {}^g\mu = \mu'^h \cdot \rho_h^{-1} \cdot {}^g\mu$ for all g and each fixed h; while the δ_g satisfy the requisite identity if and only if ${}^h\mu' \cdot {}^g\mu = \mu'^h \cdot {}^{gh^{-1}}\mu^h$. But we compute:

$$\rho_h^{-1} \cdot {}^g\mu = \rho^h_{h^{-1}} \cdot \rho_g \cdot \mu^g = (\rho_{h^{-1}} \cdot \rho_g^{h^{-1}})^h \cdot \mu^g = \rho^h_{gh^{-1}} \cdot \mu^g = {}^{gh^{-1}}\mu^h.$$

(ii). We assume the latter case, the former one being trivial. By naturality of $\delta_{gh^{-1}}$

$$\begin{array}{ccc} \underline{g} & \xrightarrow{\delta^h_{gh^{-1}}} & \underline{h} \\ {\scriptstyle {}^{gh^{-1}}\delta_h}\downarrow & & \downarrow{\scriptstyle \delta_h} \\ \underline{gh^{-1}} & \xrightarrow[\delta_{gh^{-1}}]{} & 1 \end{array}$$

commutes. Thus

$$\delta^h_{gh^{-1}} = \delta_h^{-1} \cdot \delta_g = {}^h\delta_{h^{-1}} \cdot \delta_g = {}^h(\delta_{h^{-1}} \cdot {}^{h^{-1}}\delta_g) = {}^h\delta_{h^{-1}g}. \quad \square$$

In other words, in either case of Corollary 7.10, a unit of stable reflection has a retraction compatible with the companion cotransversal of the identity transversal for insertion of the stable subcategory.

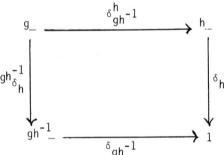

8. G-COTRIPLEABILITY

Much of the remaining memoir has to do with G-cotripleability in one form or another (including when G = 1). In the sense that there is a well-established theory of monads in 2-categories -- see Kelly-Street [20] for a condensed account -- the term "G-cotripleable" is already meaningful. Ergo, it arises from the notion of a comonad generated by an adjunction in G-Cat. However, given such a comonad, say

$$\Sigma = (MN, \rho, M\lambda N) \tag{1}$$

we will be more interested in various properties of M pertaining to certain of its factorizations by G-functors than in the comonads themselves.

We fix, for the entire section, an adjunction

$$(M,N;\lambda,\rho)\colon \mathcal{P} \to \mathcal{Q} \tag{2}$$

in G-Cat generating the comonad Σ in (1), where the notation is borrowed from [22] (i.e., M: $\mathcal{P} \to \mathcal{Q}$ is left adjoint to N with unit λ and counit ρ). We recall that a coalgebra for Σ is an ordered pair (Q,q) with underlying object $Q \in \mathcal{Q}$ and structure map $q \in \operatorname{arr}\mathcal{Q}$ such that the diagram

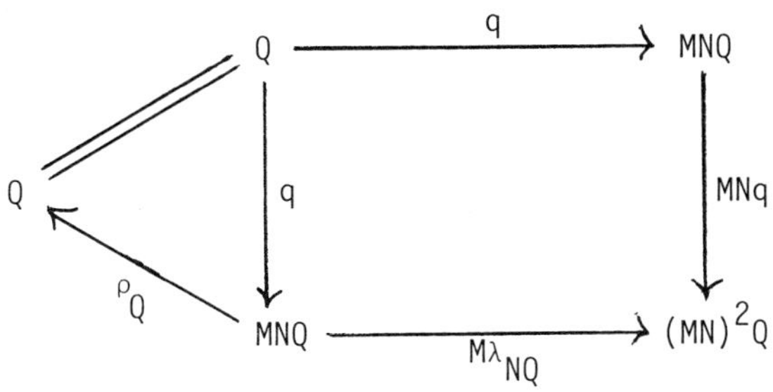

commutes -- the square being the associated law and the triangle the unit law. We use the term *M-coalgebra*. A morphism f: (Q,q) → (Q',q') of M-coalgebras is an $f \in \mathcal{Q}(Q,Q')$ such that $MNf \circ q = q' \circ f$. The resultant category $\mathcal{Q}_M$ of M-coalgebras is a G-category via ${}^g(Q,q) = ({}^gQ, {}^gq)$, where the action of G on arrows is its action on arrows of $\mathcal{Q}$. By the *standard factorization of M* we mean the factorization

$$P \xrightarrow{M} \mathcal{Q} = P \xrightarrow{\hat{M}} \mathcal{Q}_M \xrightarrow{U_M} \mathcal{Q}$$

where the underlying functor U_M and the comparison $\hat{M}$ are the G-functors as described by

$$U_M:\ \begin{array}{ccc}(Q,q) & \longmapsto & Q\\ \downarrow f & & \downarrow f\\ (Q',q') & \longmapsto & Q'\end{array} \qquad\qquad \hat{M}:\ \begin{array}{ccc} P & \longmapsto & (MP, M\lambda_P)\\ \downarrow p & & \downarrow Mp \\ P' & \longmapsto & (MP', M\lambda_{P'})\end{array} \quad .$$

Notationwise, given a G-adjunction with M a right, instead of a left, G-adjoint, we replace $\mathcal{Q}_M$, U_M and $\hat{M}$ by, respectively, $\mathcal{Q}^M$, U^M and $\check{M}$. We will justify our terminology and notation later in the section -- see Proposition 8.13.

Of course, if (2) is an adjunction in Trans-Cat, Σ is a comonad in Trans-Cat. Furthermore, assuming M to be transversaled by μ and N by ν, $\mathcal{Q}_M$ is a G-category with G operating on coalgebras (Q,q) by ${}^g(Q,q) = ({}^gQ, (\mu\nu)_g(Q) \circ {}^gq)$; and then U_M is trivially transversaled and $\hat{M}$ is transversaled by $\hat{\mu}$, where $\hat{\mu}_g$ at $P \varepsilon \mathcal{P}$ is just $\mu_g(P)$. We speak no more in the section of Trans-Cat, it being for the most part not difficult to extrapolate definitions and results there from the G-Cat case.

Continuing, we say that M is *G-cotripleable*, *injectively G-cotripleable*, or *G-comonadic* if $\hat{M}$ is respectively a G-equivalence, an injective G-equivalence, or a (necessarily G-) isomorphism. *G-adjoint tripleable* functors are both G-cotripleable and G-tripleable.

Now, by the theory in Cat, there is a (necessarily unique) adjunction $\mathcal{Q}_M \to \mathcal{Q}$ of the form $(U_M, \hat{M}N; \hat{\lambda}, \rho)$; and it too defines the comonad Σ. Thereto, inasmuch as ρ is G-natural, the adjunction is a G-adjunction. In particular, as U_M is well known to be comonadic, U_M is G-comonadic. The importance of this resides in

THEOREM 8.1. G-comonadic functors create G-colimits. ▫

It results immediately that G-comonadic functors create stable colimits (cf. 1.7). The proof of Theorem 8.1 is similiar to the next proof given -- albeit the former is easier than the latter, Bar-Wells' proof of [2, Theorem 1, p. 117] affords an alternative.

The gist of the following result is that any G-limit of a given type preserved by a G-functor having a right G-adjoint is created by its underlying G-comonadic functor.

THEOREM 8.2. If Z is a G-diagram in $\mathcal{Q}_M$ such that MN and $(MN)^2$ both preserve G-limits of $U_M Z$, then U_M creates G-limits for Z.

Proof. Let Θ be a G-limit of $U_M Z$ and, for each $j \in \mathrm{dom}Z$, set $Z_j = (Q_j, q_j)$. Then the $q_j\Theta_j$ are clearly the components of a G-cone to the base $MNU_M Z$. Thus, since $MN\Theta$ is a G-limit, there is a unique stable arrow q of $\mathcal{Q}$ making the upper square in the diagram

$$\begin{array}{ccc} Q & \xrightarrow{\Theta_j} & Q_j \\ {\scriptstyle q}\downarrow & & \downarrow{\scriptstyle q_j} \\ MNQ & \xrightarrow[MN\Theta_j]{} & MNQ_j \\ {\scriptstyle \rho_Q}\downarrow & & \downarrow{\scriptstyle \rho_{Q_j}} \\ Q & \xrightarrow[\Theta_j]{} & Q_j \end{array} \qquad Q = \lim_G U_M Z$$

commute for all j, whereas the lower square commutes for all j by naturality of ρ. In particular, since Θ is a G-limit and the $\rho_{Q_j} q_j \Theta_j$ define a G-cone to $U_M Z$, commutativity of the perimeter of the diagram for all j implies the unit law for the pair (Q,q). But also, by naturality of $M\lambda$

$$\begin{array}{ccc} MNQ & \xrightarrow{M\lambda_{NQ}} & (MN)^2 Q \\ {\scriptstyle MN\Theta_j}\downarrow & & \downarrow{\scriptstyle (MN)^2\Theta_j} \\ MNQ_j & \xrightarrow[M\lambda_{NQ_j}]{} & (MN)^2 Q_j \end{array}$$

commutes. Consequently

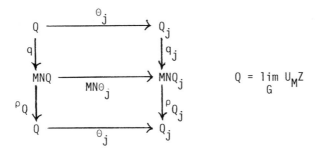

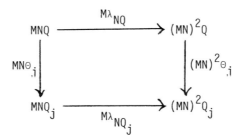

$$(MN)^2\Theta_j \circ MNq \circ q = MNq_j \circ MN\Theta_j \circ q = MNq_j \circ q_j\Theta_j$$

$$= M\lambda_{NQ_j} \circ q_j\Theta_j = M\lambda_{NQ_j} \circ MN\Theta_j \circ q = (MN)^2\Theta_j \circ M\lambda_{NQ} \circ q.$$

Thus, since $(MN)^2\Theta$ is a G-limit and $MNq \circ q$ and $M\lambda_{NQ} \circ q$ are stable, the associative law is valid. So (Q,q) is a stable M-coalgebra and μ: $(Q,q) \to Z$, defined by $\mu_j = \Theta_j$, is a G-cone. Evidently, $U_M\mu = \Theta$ and, thereby, uniqueness of μ is plain.

To show μ is a G-limit, consider another G-cone, say μ': $(Q',q') \to Z$. Then, as Θ is a G-limit of U_MZ, $U_M\mu' = \Theta f$ for a unique $f \in \mathrm{stab}\mathcal{Q}(Q',Q)$. Clearly it suffices that f be a coalgebra morphism $(Q',q') \to (Q,q)$; and this follows upon composing the $MN\Theta_j$ with the stable arrows $MNf \circ q'$ and qf. □

Since, by Theorem 6.12, N is G-continuous:

COROLLARY 8.3. If M is G-cotripleable, then $\mathcal{P}$ is as G-complete as $\mathcal{Q}$ is G-complete and M is G-contiuous. □

Theorem 8.2 has a partial converse; namely, if M is G-cotripleable then M preserves any G-limit of a given type which exist in $\mathcal{Q}$ and is created by U_M. Thus, for example, a G-comonadic functor with G-complete codomain preserves precisely the G-limits it creates.

The next two results are both, in part, consequences of the fact that comonadic functors create isomorphisms (hence reflect identity arrows).

PROPOSITION 8.4. A G-cotripleable functor is injectively G-cotripleable exactly when it reflects identity arrows and G-comonadic exactly when it creates isomorphisms.

Proof. See the comment above 3.4 concerning injectivity on objects. □

THEOREM 8.5. Let F: $\mathcal{X} \to \mathcal{Y}$ be a G-functor.

(i) If $\mathcal{X}$ is hereditarily stably closed and F is cotripleable, then F is G-cotripleable.

(ii) If F is G-cotripleable and $\mathcal{Y}$ is stably closed (resp. hereditarily stably closed) then $\mathcal{X}$ is stably closed (resp. hereditarily stably closed).

Proof. (ii) follows using 2.11 and (i) follows upon applying 3.12 twice and using 3.3. □

As promised in the preceding section:

THEOREM 8.6. If G operates without fixed points on objJ and C is a G-category, then the diagonal G-functor $C \to C^{\dot{J}}$ is G-tripleable if and only if every diagram in C over J has a colimit and C is hereditarily stably closed. In particular, the hereditarily stably closed G-categories C with G-indexed coproducts are precisely those for which the diagonal $C \to C^{\dot{G}}$ is G-tripleable.

Proof. By 7.4 and the preceding theorem (both parts), it is yet to show that $\Delta\colon C \to C^{\dot{J}}$ is tripleable whenever it has a left adjoint. This is an essentially straightforward consequence of PTT. □

We comment that, from the second statement in this result, one can construct at will examples of tripleable G-functors that are not G-tripleable (take C = suitable trivial G-category). Thus one should have criteria for a G-functor to be G-tripleable that do not require of the domain that it be hereditarily stably closed. These are afforded by the next two results.

PROPOSITION 8.7. M is G-cotripleable when and only when $\mathrm{stab}_H M\colon \mathrm{stab}_H \mathcal{P} \to \mathrm{stab}_H \mathcal{Q}$ is cotripleable for every subgroup H of G.

Proof. Immediate by 4.8 and 3.3. □

We say that a parallel pair of stable arrows in a G-category X has a *stably split equalizer* if it has a split equalizer in stab X. Notice that if

$$\cdot \xrightarrow{\ e\ } \cdot \underset{v}{\overset{u}{\rightrightarrows}} \cdot$$

is a stably split fork, then e is a stable equalizer of u and v.

THEOREM 8.8 (PTT for G-categories). M is G-cotripleable if and only if, for every subgroup H of G, $\mathcal{P}$ has, and M preserves and reflects, H-stable equalizers for any H-stable parallel pair x,y in $\mathcal{P}$ for which Mx,My has an H-stably split equalizer in $\mathcal{Q}$.

Proof. Given the preceeding proposition, this follows by ordinary PTT, 1.2 and 1.3. □

As a routine consequence (the first part uses 8.1 too):

COROLLARY 8.9. Suppose M is G-cotripleable.

(i) M is G-adjoint tripleable if and only if M has a left G-adjoint.

(ii) If E is a G-equivalence, EM is G-cotripleable whenever EM is defined and ME is G-cotripleable whenever ME is defined. □

The rest of the section is concerned with an analysis of the relationship between the standard factorization of M and other factorizations of M by G-functors.

Now, as is checked from the relevant triangle identities, $({}^gM, {}^{g^{-1}}N; \lambda, \rho)$: $\mathcal{P} \to \mathcal{Q}$ is an adjunction for each $g \in G$. Not only do they all generate the comonad Σ, but the two factorizations

$$\mathcal{P} \xrightarrow{\widehat{{}^gM}} \mathcal{Q}_{{}^gM} \xrightarrow{U_{{}^gM}} \mathcal{P} \qquad\qquad \mathcal{P} \xrightarrow{{}^g\hat{M}} \mathcal{Q}_M \xrightarrow{{}^gU_M} \mathcal{Q}$$

of ${}^gM$ are always identical. In particular, ${}^gM$ is comonadic if M is.

The foregoing constitutes a principle which can at times be used in extending known results concerning adjunctions to G-adjunctions. A case in point is the following version of a useful technical result of Mac Lane [22, Lemma, p. 149].

LEMMA 8.10. If $(M', N'; \lambda', \rho')$: $\mathcal{P}' \to \mathcal{Q}$ is a G-adjunction generating the comonad Σ and M' is comonadic, then the diagram

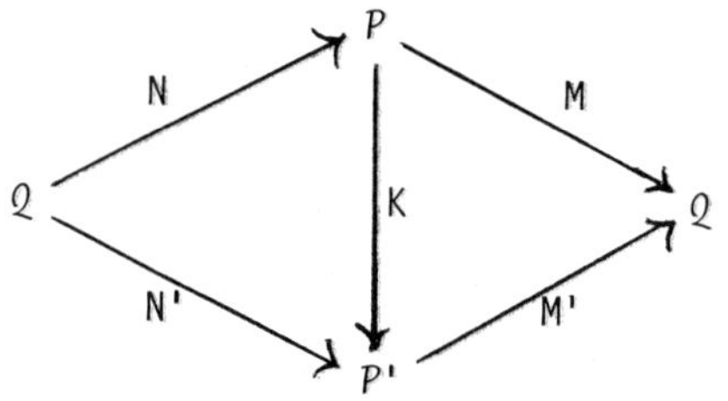

is rendered commutative by a unique G-functor K (i.e., K is a "G-comparison").

Proof. As we just saw $({}^{g^{-1}}M', {}^gN'; \lambda', \rho')$: $\mathcal{P}' \to \mathcal{Q}$ is an adjunction generating Σ and ${}^{g^{-1}}M'$ is comonadic. Hence, by the cited result of Mac Lane, there is a unique functor $K(g)$: $\mathcal{P} \to \mathcal{P}'$ such that ${}^{g^{-1}}M'K(g) = M$ and $K(g)N = {}^gN'$. In particular $K = K(1)$ is unique with $M'K = M$ and $KN = N'$. But ${}^{g^{-1}}M'{}^gK = M'K = M$ and ${}^gKN = {}^g(KN) = {}^gN'$. Thus ${}^gK = K(g)$. But also, ${}^{g^{-1}}M'K^g = {}^{g^{-1}}(M'K)^g =$

$g^{-1}M^g = M$ and $K^gN = (KN)^g = N'^g = {}^gN'$. Thus $K^g = K(g)$. □

COROLLARY 8.11. Let

$$P \xrightarrow{E} X \xrightarrow{F} \mathcal{Q} \qquad\qquad P \xrightarrow{E'} X' \xrightarrow{F'} \mathcal{Q}$$

be factorizations of M by G-functors with F' G-comonadic. If there are G-adjunctions $X \to \mathcal{Q}$ and $X' \to \mathcal{Q}$ of the respective forms (F,EN;–,–) and (F',E'N;–,–) generating the comonad Σ, then the diagram

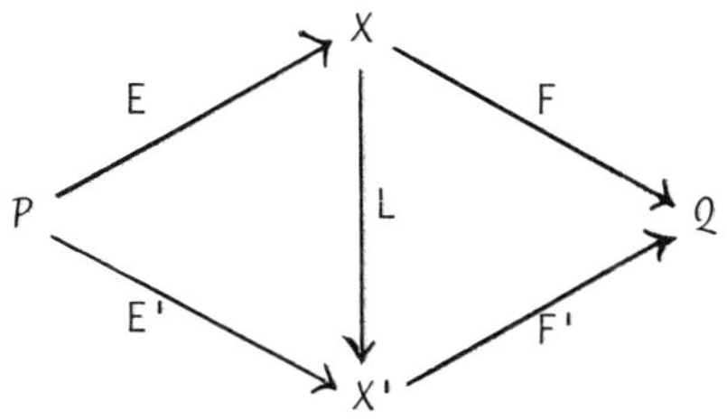

commutes for a unique G-functor L.

Proof. According to the lemma, there is a unique G-functor $L\colon X \to X'$ such that $F'L = F$ and $LEN = E'N$ and a unique G-functor $K\colon P \to X'$ such that $F'K = M$ and $KN = E'N$. But, since $F'LE = FE = M$, $LE = K$ and, since $F'E' = M$, $E' = K$. Therefore $LE = E'$. Lastly, if L_1 is a G-functor satisfying $F'L_1 = F$ and $L_1E = E'$ then, inasmuch as $L_1EN = E'N$, we infer that $L_1 = L$. □

Naturally, if F is G-comonadic too, L is an isomorphism. In fact, one can show L factors as $X \xrightarrow{\hat{F}} \mathcal{Q}_F \longrightarrow X'$ for some G-isomorphism $\mathcal{Q}_F \longrightarrow X'$. Also, the corollary has the consequence that if M is G-comonadic

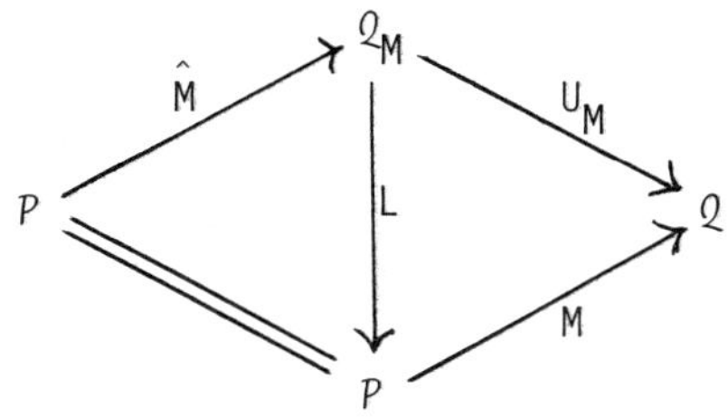

commutes for a unique G-isomorphism L. In this case we identify $\mathcal{Q}_M$ with P via L, thereby identifying U_M with M and $\hat{M}$ with 1_P. Similiarly, if M is a G-equivalence, U_M is the G-isomorphism $L\colon \mathcal{Q}_M \to \mathcal{Q}$ satisfying $L\hat{M} = M$. Here we identify $\mathcal{Q}_M$ with $\mathcal{Q}$ via U_M, thus identifying U_M with $1_{\mathcal{Q}}$ and $\hat{M}$ with M.

Due to the last result, the next is of interest.

LEMMA 8.12. If $\mathcal{P} \xrightarrow{E} \mathcal{X} \xrightarrow{F} \mathcal{Q}$ is a factorization of M by G-functors with E a G-equivalence, there is a G-adjunction $(F,EN;\Psi,\rho)\colon \mathcal{X} \to \mathcal{Q}$ generating the comonad Σ.

Proof. Choose a G-adjoint equivalence, say $(T,E;\omega,\zeta)\colon \mathcal{X} \to \mathcal{P}$. We claim that $\Psi = ENF\omega^{-1} \cdot E\lambda T \cdot \omega$ works. For this, observe that the sequence of canonical natural isomorphisms

$$\mathcal{Q}(FX,Q) \xrightarrow{(F\omega_X^{-1})^*} \mathcal{Q}(MTX,Q) \longrightarrow \mathcal{P}(TX,NQ) \longrightarrow \mathcal{X}(X,ENQ)$$

exhibits F as left adjoint of EN. Setting $Q = FX$ and evaluating at 1_{FX} we get that the unit is the one claimed. Setting $X = ENQ$ and evaluating at 1_{ENQ}, since

$$M\zeta N \cdot F\omega EN = M\zeta N \cdot FE\zeta^{-1}N = M\zeta N \cdot M\zeta^{-1}N = 1$$

we get that ρ is the counit. But, since for the indicated choice of Ψ

$$\Psi EN = ENF\omega^{-1}EN \cdot E\lambda TEN \cdot \omega EN = ENM\zeta N \cdot E\lambda TEN \cdot E\zeta^{-1}N$$

by naturality of $M\lambda$ we have

$$F\Psi EN = MNM\zeta N \cdot MNM\zeta^{-1}N \cdot M\lambda N = M\lambda N . \quad \square$$

In the same vein we prove, for the record:

PROPOSITION 8.13. Let $(M,N';\lambda',\rho')\colon \mathcal{P} \to \mathcal{Q}$ be an adjunction in G-Cat and denote the comonad it generates by Σ'. Then

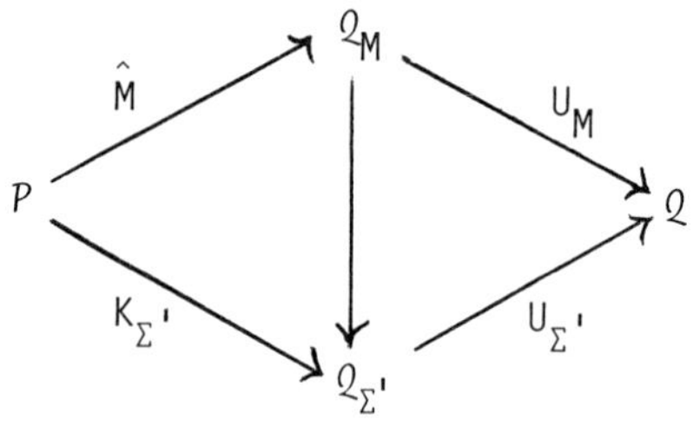

commutes for a unique G-isomorphism (the vertical arrow), the lower factorization of M being its standard factorization relative to Σ'.

Proof. Consider the composite natural isomorphism

$$\mathcal{Q}(U_{\Sigma'}-,-) \longrightarrow \mathcal{Q}_{\Sigma'}(-,K_{\Sigma'}N'-) \xrightarrow{(K_{\Sigma'}\eta)_*} \mathcal{Q}_{\Sigma'}(-,K_{\Sigma'}N-)$$

where $\eta\colon N' \to N$ is the G-natural isomorphism subject to $\rho \cdot M\eta = \rho'$ and

$\eta M \cdot \lambda' = \lambda$. By inspection of this composite we find $(U_{\Sigma'}, K_{\Sigma'}N; K_{\Sigma'}\eta U_{\Sigma'} \cdot \lambda'_{\Sigma'}, \rho): \mathcal{Q}_{\Sigma'} \to \mathcal{Q}$ is a G-adjunction, where $\lambda'_{\Sigma'}$ is the unit of the G-adjunction $(U_{\Sigma'}, K_{\Sigma'}N'; \lambda'_{\Sigma'}, \rho'): \mathcal{Q}_{\Sigma'} \to \mathcal{Q}$ determined by the given G-adjunction $(M,N';\lambda',\rho')$. But

$$U_{\Sigma'}(K_{\Sigma'}\eta U_{\Sigma'} \cdot \lambda'_{\Sigma'})K_{\Sigma'}N = U_{\Sigma'}K_{\Sigma'}\eta U_{\Sigma'}K_{\Sigma'}N \cdot U_{\Sigma'}\lambda'_{\Sigma'}K_{\Sigma'}N$$

$$= M\eta MN \cdot U_{\Sigma'}K_{\Sigma'}\lambda' N = M\eta MN \cdot M\lambda' N = M\eta MN \cdot M\eta^{-1}MN \cdot M\lambda N = M\lambda N.$$

So the rest follows by 8.11. □

THEOREM 8.14. Consider the diagram of G-functors

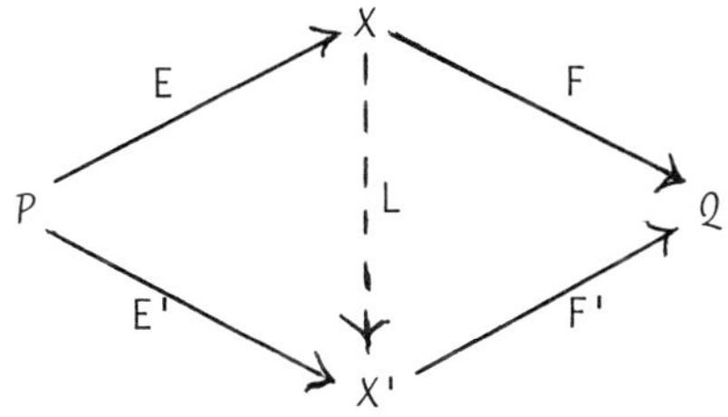

(ignore the dashed arrow) where E is a G-equivalence and F' creates isomorphisms. Given a G-natural isomorphism $\Psi: FE \to F'E'$, there is a unique G-functor L as pictured, such that the triangle on the right commutes and the triangle on the left commutes up to a (necessarily unique) G-natural isomorphism $\phi: LE \to E'$ with $F'\phi = \Psi$. Thus if $FE = F'E'$ the diagram commutes for a unique G-functor L.

Proof. Let $(E,R;\omega,\zeta): \mathcal{P} \to \mathcal{X}$ be a G-adjoint equivalence and consider the G-natural isomorphism $\eta = F\zeta \cdot \Psi^{-1}R: F'E'R \to F$. Since F' creates isomorphisms, for each $X \in \mathcal{X}$ there is a unique arrow of the form $\mu_X: E'RX \to LX$ in $\mathcal{X}'$ such that $F'\mu_X = \eta_X$; and each μ_X is an isomorphism. It follows that L is object function of a unique G-functor $L: \mathcal{X} \to \mathcal{X}'$ for which the μ_X are the components of a G-natural isomorphism $\mu: E'R \to L$ with $F'\mu = \eta$. In particular, $F'L = F$.

Let ϕ be the G-natural isomorphism $E'\omega^{-1} \cdot \mu^{-1}E: LE \to E'$. Then

$$F'\phi = F'E'\omega^{-1} \cdot F'\mu^{-1}E = F'E'\omega^{-1} \cdot \eta^{-1}E = F'E'\omega^{-1} \cdot \Psi RE \cdot F\zeta^{-1}E$$

$$= F'E'\omega^{-1} \cdot \Psi RE \cdot FE\omega = F'E'\omega^{-1} \cdot F'E'\omega \cdot \Psi = \Psi$$

where the next to last step is naturality of Ψ.

Having established existence, suppose that $L_1: X \to X'$ is another G-functor satisfying $F'L_1 = F$ for which there is a G-natural isomorphism $\phi_1: L_1E \to E'$ such that $F'\phi_1 = \Psi$. Then, from the fact that F' reflects identity arrows, we deduce that $\phi_1 = \phi$; hence that $L_1E = LE$. But then $\sigma = L\zeta \cdot L_1\zeta^{-1}$ is a natural transformation $L_1 \to L$ satisfying

$$F'\sigma = F'L\zeta \cdot F'L_1\zeta^{-1} = F\zeta \cdot F\zeta^{-1} = 1.$$

Consequently $L_1 = L$. □

This result, in almost full generality, will be a major factor in Section 12. Meanwhile, the special case $FE = F'E'$ implies the next result -- except that validity of the sufficient condition for G-cotripleability in part (ii) requires Corollary 8.9(ii) (or 8.11 and 8.12).

COROLLARY 8.15.(i) If M is G-cotripleable and $\mathcal{P} \longrightarrow X \longrightarrow \mathcal{Q}$ is any factorization of M by G-functors such that $X \to \mathcal{Q}$ creates isomorphisms, there exists a unique G-functor L: $\mathcal{Q}_M \to X$ rendering commutative the diagram

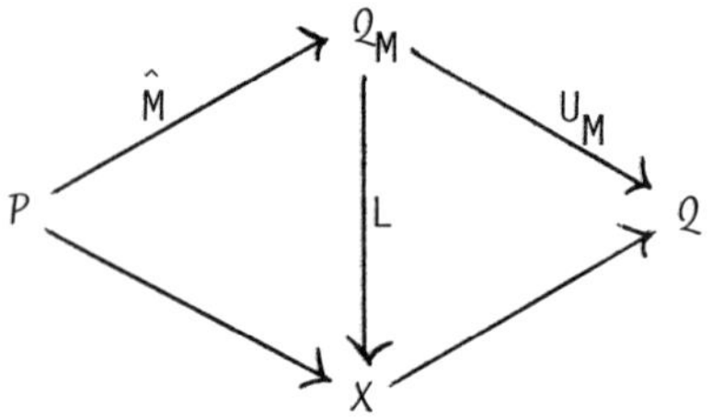

Moreover, L is an isomorphism exactly when $\mathcal{P} \to X$ is a G-equivalence.

(ii) A functor is G-cotripleable if and only if it is the composite of a G-comonadic functor with a G-equivalence; it is G-adjoint tripleable if and only if it is the composite of a G-comonadic-G-monadic functor with a G-equivalence. These factorizations are unique apart from a unique G-isomorphism. □

One deduces at once that if FE is G-cotripleable, E is a G-equivalence and F is an isomorphism creating G-functor, then F is G-comonadic. Alternatively, this follows by Corollary 8.9 and Proposition 8.4.

9. THE STANDARD FACTORIZATION OF INSERTION

In this section we study the standard factorization (see § 8) of insertion, I, of the reflective stable subcategory of a G-category C. We show there is a canonical faithful isomorphism-reflecting functor from the category C^I of I-algebras to the category stab C^G of systems of isomorphisms. Indeed, we give necessary and sufficient conditions for the factorization

$$\text{stab } C \xrightarrow{\underline{\Delta}} \text{stab } C^G \xrightarrow{E} C$$

of I to be, up to unique isomorphism, the standard factorization, where $\underline{\Delta}$ is the stable diagonal and E is the restriction of evaluation at the identity of G. When the stable subcategory is stably reflective, we show that the sufficient conditions are fulfilled. Thus, for a G-category with stable reflections, insertion of the stable subcatgeory is tripleable if and only if the category is stably closed -- cf. Theorem 7.1.

The above notation will be retained throughout the section. In addition, the reflector for reflection of C in stab C will be denoted by R and the unit and counit of reflection by μ and ν, respectively. At times we shall regard R as an endofunctor on C.

THEOREM 9.1. The diagram

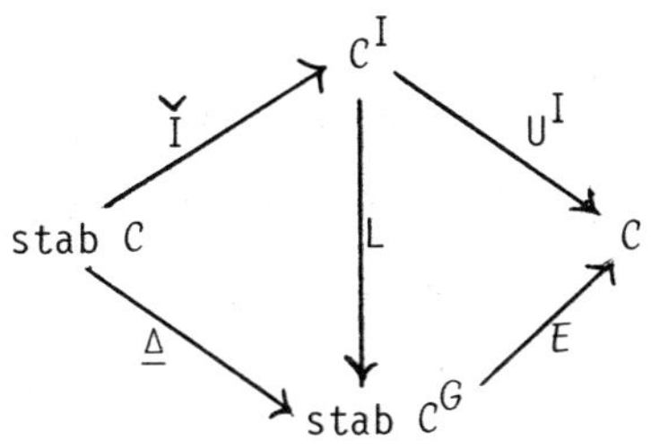

(1)

commutes for a unique functor L; and this L is faithful and reflects isomorphisms.

Proof. Supposing (C,c) to be an I-algebra, the diagram

$$\begin{array}{ccccc}
RC & \xrightarrow{\mu_{RC}} & R^2C & \xrightarrow{\nu_{RC}} & RC \\
{\scriptstyle c}\downarrow & & \Big| {\scriptstyle Rc} & & \downarrow{\scriptstyle c} \\
C & \xrightarrow[\mu_C]{} & RC & \xrightarrow[c]{} & C
\end{array}$$

is a split coequalizer. This entails, for any $g\varepsilon G$, that

$$c = c \circ {}^g1_{RC} = c \circ {}^g(\nu_{RC}\mu_{RC}) = c\nu_{RC} \circ {}^g\mu_{RC}$$
$$= c \circ Rc \circ {}^g\mu_{RC} = c \circ {}^g(Rc \circ \mu_{RC}) = c \circ {}^g(\mu_C c)$$

demonstrating that $c = c^g(\mu_C c)$ for all g.

We claim that the G-indexed family of arrows

$$\phi_g = c^g\mu_C : {}^gC \to C$$

forms a system of isomorphisms. But

$$\phi_g{}^g\phi_h = c^g\mu_C{}^gc^{gh}\mu_C = c^g(\mu_C c)^{gh}\mu_C = c^{gh}\mu_C = \phi_{gh}.$$

This, and the fact that $\phi_1 = c\mu_C = 1$, verifies our claim.

Now if $f:(C,c) \to (D,d)$ is an algebra morphism, with $\phi_g = c^g\mu_C$ and $\psi_g = d^g\mu_D$, we have

$$\psi_g{}^gf = d^g\mu_D{}^gf = d^g\mu_D{}^gf^g(c\mu_C) = d^g\mu_D{}^g(fc)^g\mu_C$$
$$= d^g\mu_D{}^g(d \circ Rf)^g\mu_C = d^g(\mu_D d) \circ {}^gRf \circ {}^g\mu_C$$
$$= d \circ Rf \circ {}^g\mu_C = fc^g\mu_C = f\phi_g .$$

Thus L defined by

$$L(C,c) = \{c^g\mu_C\}_{g\varepsilon G} \qquad\qquad Lf = f$$

is a functor. Manifestly, L is faithful and reflects isomorphisms.

If $C \varepsilon$ stab $\mathcal{C}$, since

$$(L\check{I}C)_g = \nu_C{}^g\mu_C = {}^g(\nu_C\mu_C) = {}^g1_C = 1_C = (\underline{\Delta}C)_g$$

it follows easily that $L\check{I} = \underline{\Delta}$. It is evident that $EL = U^I$.

It remains to establish uniqueness of L. So let $L_1:\mathcal{C}^I \to \text{stab } \mathcal{C}^G$ be such that the diagram (1) commutes with L_1 in place of L, and put $L_1(C,c) = \phi$. But c is a morphism of algebras $(RC,\nu_{RC}) \to (C,c)$, so that, by commutativity of the left triangle in (1), L_1c is an arrow $\underline{\Delta}RC \to \phi$. This means that $\phi_g{}^gc = c$. But then

$$c^g\mu_C = \phi_g{}^gc^g\mu_C = \phi_g{}^g(c\mu_C) = \phi_g{}^g1_C = \phi_g .$$

Therefore L_1 agrees with L on objects. Commutativity of the triangle on the right ensures L_1 agrees with L on arrows. □

We know of no noncomputational proof of Theorem 9.1. However, with various additional hypotheses, the theorem follows by earlier results. For instance, if I is tripleable, existence and uniqueness of L, together with the fact that L is an isomorphism if and only if C is stably closed, is an immediate consequence of Theorem 8.15(i) abetted by Theorem 2.1 and Corollary 2.8. On the other hand, if C is stably closed, the existence of a unique isomorphism L making (1) commute follows by Corollaries 2.8 and 8.11, Lemma 8.12, and Theorem 2.1.

We recall that a, say, right adjoint S is said to be of descent type if its comparison $\check{S}$ is fully faithful.

LEMMA 9.2. I is of descent type.

Proof. Let C,D ε stab C and suppose that $\gamma \in C(C,D)$ makes the square

$$\begin{array}{ccc} RC & \xrightarrow{R\gamma} & RD \\ {\scriptstyle \nu_C}\downarrow & & \downarrow{\scriptstyle \nu_D} \\ C & \xrightarrow[\gamma]{} & D \end{array}$$

commute. One infers immediately that ${}^g\gamma\nu_C = \gamma\nu_C$ for any g. Then, multiplying this equation on the right by μ_C, one gets that ${}^g\gamma = \gamma$. So $\check{I}$ is full. However, since I is faithful, $\check{I}$ is automatically faithful. □

THEOREM 9.3. I is tripleable precisely when the functor L of Theorem 9.1 factors through the full subcategory of stabilizing systems of isomorphisms in C, in which case corestriction of L to this full subcategory is a surjective equivalence.

Proof. We claim that if the system of isomorphism L(C,c) is stabilizing, then the I-algebra (C,c) is isomorphic to a free I-algebra. Notice that validity of this claim proves the sufficiency of the assertion concerning tripleability of I, by 9.2. Now, the assumption L(C,c) is stabilizing means there is a stable D and isomorphism γ: C $\rightarrow$ D such that $c^g\mu_C = \gamma^{-1}{}^g\gamma$.

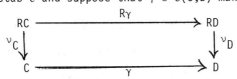

In particular, not only the upper square, but the perimeter in the diagram

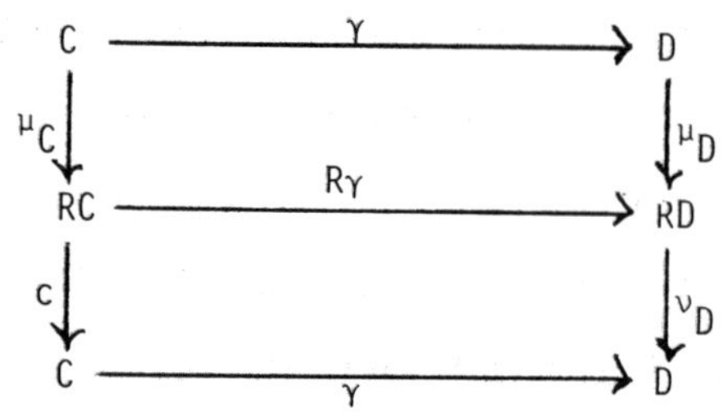

commute. But since, by the proof of 9.1, $c = c^{g}(\mu_C c)$

$$\gamma c = \gamma c^{g}\mu_C{}^{g}c = \gamma\gamma^{-1}{}^{g}\gamma{}^{g}c = {}^{g}(\gamma c)$$

that is, γc is stable. Thus, since $\nu_D R\gamma$ is stable too, the lower square in the diagram commutes. Consequently (C,c) is isomorphic to the free I-algebra (D,ν_D), verifying our claim.

Next, assuming I to be tripleable, it is plain from the foregoing that each $L(C,c)$ is a stabilizing system of isomorphisms. Thus, by full faithfulness of $\underline{\Delta}$, we are left with showing that the prescribed corestriction is surjective on objects. For this, let ϕ be a stabilizing system of isomorphisms at, say, C. Then, as before, $\phi_g = \gamma^{-1}{}^{g}\gamma$; so, in particular

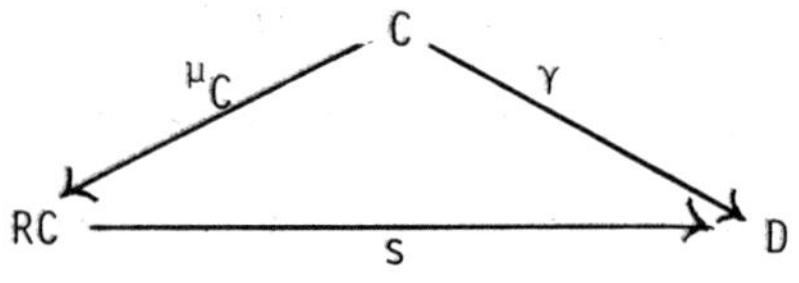

commutes for a stable s. We deduce, as before

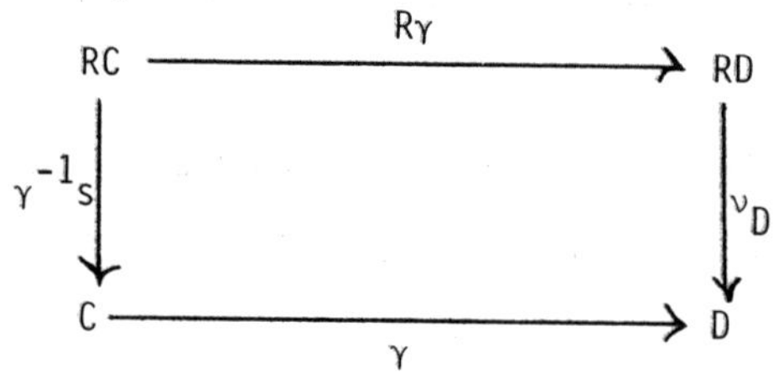

is commutative, therefrom we conclude $(C,\gamma^{-1}s)$ is an algebra (see 10.3). Moreover, since

$$\gamma^{-1}s\,g_{\mu_C} = \gamma^{-1}g(s\mu_C) = \gamma^{-1}g_{\gamma} = \phi_g$$

$L(C,\gamma^{-1}s) = \phi$. □

By the proof of this result and Corollary 2.4, one has (cf. 10.5)

COROLLARY 9.4. Let $C \varepsilon \mathcal{C}$. There exists a stabilizing system of isomorphisms at C if and only if C is the underlying object of an I-algebra isomorphic to a free I-algebra. Thus, if I is tripleable, C is the underlying object of an I-algebra if and only if C is isomorphic to a stable object. □

The next result, among other conditions equivalent to L being an isomorphism, lists the exact assumptions on the unit μ of reflection required therefor. This result, allied with the one coming afterwards, illustrates the utility of studying stable reflections versus ordinary reflections of a G-category in its stable subcategory.

THEOREM 9.5. Referring to the diagram (1), the following are equivalent statements.

(i) L is an isomorphism.

(ii) E has a left adjoint which factors through $\underline{\Delta}$.

(iii) $\underline{\Delta}R$ is left adjoint to E with unit μ.

(iv) Given an arrow γ: $C \to D$ in $\mathcal{C}$ and a system of isomorphisms Ψ at D, there is a unique arrow f: $RC \to D$ of $\mathcal{C}$ rendering commutative the diagram

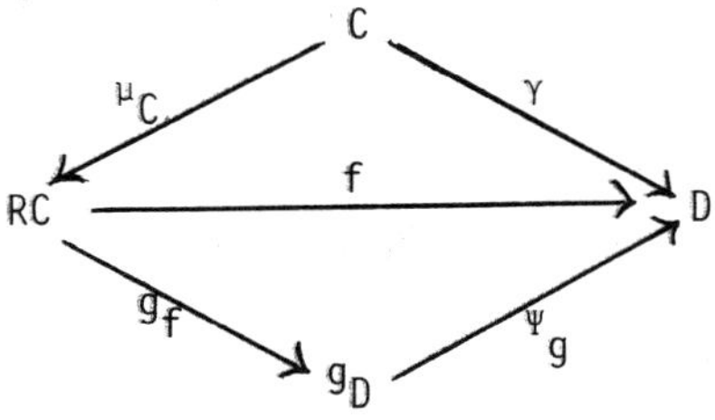

(2)

for every $g \varepsilon G$.

Proof. (i) =>(ii). Since $\check{I}R$ is left adjoint to U^I, $\underline{\Delta}R = L\check{I}R$ is left adjoint to E.

(ii) =>(iii). Let $\underline{\Delta}T$ be left adjoint to E with unit ω, where T: $\mathcal{C} \to$ stab $\mathcal{C}$. Then, by the proof of 7.2, T is left adjoint to I with the same unit ω. So there is a natural isomorphism, say η: $R \to T$, such that $I\eta \cdot \mu = \omega$.

Thus, $\underline{\Delta}\eta\colon \underline{\Delta}R \to \underline{\Delta}T$ is a natural isomorphism such that $E\underline{\Delta}\eta \cdot \mu = \omega$.

(iii)<=>(iv). The statement in (iv) says precisely that $(\underline{\Delta}RC,\mu_C)$ is an initial object of the comma category $(C{\downarrow}E)$.

(iii) =>(i). Let ζ be the counit of the adjunction given. One knows that $\check{I}R$ is left adjoint to U^I with unit μ and counit $\check{\nu}$ satisfying $U^I\check{\nu}\check{I}R = I\nu R$. But, insofar as

$$\begin{array}{ccccc} C & \xrightarrow{\ \check{I}R\ } & C^I & \xrightarrow{\ U^I\ } & C \\ \| & & \big\downarrow L & & \| \\ C & \xrightarrow[\ \underline{\Delta}R\ ]{} & \text{stab } C^G & \xrightarrow[\ E\ ]{} & C \end{array}$$

commutes, it is standard that $L\check{\nu} = \zeta L$; and one gets

$$E\zeta\underline{\Delta}R = E\zeta L\check{I}R = EL\check{\nu}\check{I}R = U^I\check{\nu}\check{I}R = I\nu R.$$

Consequently L is an isomorphism by 8.11, since, by 2.1, E is monadic. □

THEOREM 9.6. If C has stable reflections, L is an isomorphism. In particular an object of C is the underlying object of some I-algebra exactly when there is a system of isomorphisms at that object.

Proof. We verify condition (iv) of the preceding result: Let $\gamma \in C(C,D)$ and $\Psi \in \text{stab } C^G$ with $\Psi 1 = D$. We may choose $d \in C(RD,D)$ such that $d^g\mu_D = \Psi_g$ for all g. Then, for every h

$$\Psi_g{}^g d^h\mu_D = {}^g({}^{g^{-1}}\Psi_g d^{g^{-1}h}\mu_D) = {}^g({}^{g^{-1}}\Psi_g\Psi_{g^{-1}h}) = \Psi_g{}^g\Psi_{g^{-1}h} = \Psi_h = d^h\mu_D.$$

Thus $\Psi_g{}^g d = d$ for all g.

Put $f = d \circ R\gamma\colon RC \to D$. Then

$$\Psi_g{}^g f = \Psi_g \circ {}^g d \circ R\gamma = d \circ R\gamma = f.$$

Also, by naturality of μ

$$f\mu_C = d \circ R\gamma \circ \mu_C = d\mu_D\gamma = \Psi_1\gamma = \gamma.$$

Thus f makes (2) commute.

Finally, suppose (2) commutes when f is replaced by $f'\colon RC \to D$. We then have

$$f'{}^g\mu_C = \Psi_g{}^g f'{}^g\mu_C = \Psi_g{}^g(f'\mu_C) = \Psi_g{}^g\gamma = d^g\mu_D{}^g\gamma = d^g(\mu_D\gamma)$$
$$= d^g(R\gamma \circ \mu_C) = d \circ R\gamma \circ {}^g\mu_C = f^g\mu_C.$$

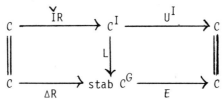

So f' = f. □

THEOREM 9.7. Any G-category with stable reflections is stably closed, provided insertion of its stable subcategory is tripleable. □

According to [9], there are non-stably closed G-categories for which the stable subcategory is an equivalent subcategory (hence reflective with tripleable insertion). Thereby, in [9], are examples of stably closed G-categories having reflective non-stably reflective stable subcategory as well as having nonreflective stable subcategory. Also, see Corollary 13.11 for an interesting application of Theorem 9.7.

10. COTRIPLEABILITY OF STABLE REFLECTORS

We call a stable idempotent of a G-category *stably split* if it is split in the stable subcategory. If every stable idempotent is stably split, we say, following Freyd [7], that the category is *stably amenable*; and if the category is H-stably amenable for every subgroup H, we call it *hereditarily stably amenable*. One should be cognizant of the fact that in a stably closed G-category every split stable idempotent is stably split. This an immediate consequence of Corollary 1.6, or of the discussion preceding Proposition 6.5. Thus, for hereditarily stably closed G-categories, the terms "amenable" and "hereditarily stably amenable" are synonymous.

In the terminology just introduced, the general result concerning G-cotripleability that follows -- pursuance of which seems worthwhile in its own right -- may be stated. Our chief result, Theorem 10.12, on cotripleability of stable reflectors will be a consequence.

THEOREM 10.1. Any left G-adjoint with hereditarily stably amenable domain and whose unit of G-adjunction has a G-natural left inverse is G-cotripleable.

This will be proved below. Meanwhile, let

$$(M,N;\lambda,\rho)\colon \mathcal{P} \to \mathcal{Q} \tag{1}$$

be an arbritrary G-adjunction. We begin with a technical lemma:

LEMMA 10.2. Let $q \in \mathrm{stab}\mathcal{Q}(Q,MNQ)$ and $p \in \mathrm{stab}\mathcal{P}(P,NQ)$.

(i)

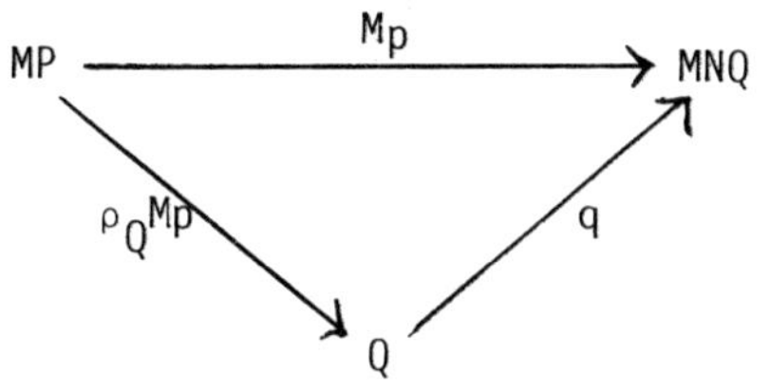

(2)

commutes if and only if

$$\begin{array}{ccc} P & \xrightarrow{\;p\;} & NQ \\ {\scriptstyle p}\downarrow & & \downarrow{\scriptstyle Nq} \\ NQ & \xrightarrow[\lambda_{NQ}]{} & NMNQ \end{array} \qquad (3)$$

commutes.

(ii) If q is a section of ρ_Q and p' a stable retraction of p, then q is the equalizer of Mpp' and 1 if and only if (2) commutes and $\rho_Q Mp$ is an isomorphism.

Proof. (i). By naturality of λ, $\lambda_{NQ}p = NMp \circ \lambda_P$ whereas, using the triangle identity $N\rho \cdot \lambda N = 1$

$$Nq \circ p = Nq \circ N\rho_Q \circ \lambda_{NQ} \circ p = Nq \circ N\rho_Q \circ NMp \circ \lambda_P = N(q\rho_Q Mp) \circ \lambda_P.$$

But (MP,λ_P) is an initial object of $(P{\downarrow}N)$.

(ii). To say that q is equalizer of Mpp' and 1 is to say that $q \circ \rho_Q \circ Mpp' = Mpp'$ and $\rho_Q \circ Mpp' \circ q = 1$. The former equation is equivalent to commutativity of (2), in which case $Mp' \circ q$ is a left inverse of $\rho_Q Mp$:

$$Mp' \circ q \circ \rho_Q \circ Mp = Mp' \circ Mp = 1.$$

Thus, when (2) commutes, the latter equation is equivalent to $\rho_Q Mp$ being an isomorphism. □

LEMMA 10.3. If $q \in \mathrm{stab}\mathcal{Q}(Q,MNQ)$ and $P \in \mathrm{stab}\,\mathcal{P}$, the following statements are equivalent.

(i) (2) commutes for some $p \in \mathrm{stab}\mathcal{P}(P,NQ)$ such that $\rho_Q Mp$ is an isomorphism.

(ii)

$$\begin{array}{ccc} MP & \xrightarrow{\;x\;} & Q \\ {\scriptstyle M\lambda_P}\downarrow & & \downarrow{\scriptstyle q} \\ MNMP & \xrightarrow[MNx]{} & MNQ \end{array}$$

commutes for some stable isomorphism x.

(iii) $(Q,q) \in \mathcal{Q}_M$ and there exists $p \in \mathrm{stab}\mathcal{P}(P,NQ)$ such that

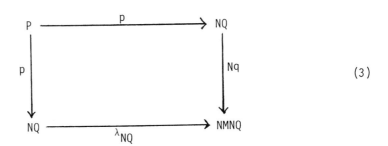

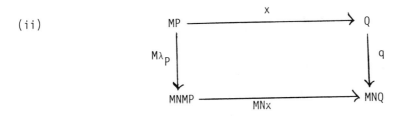

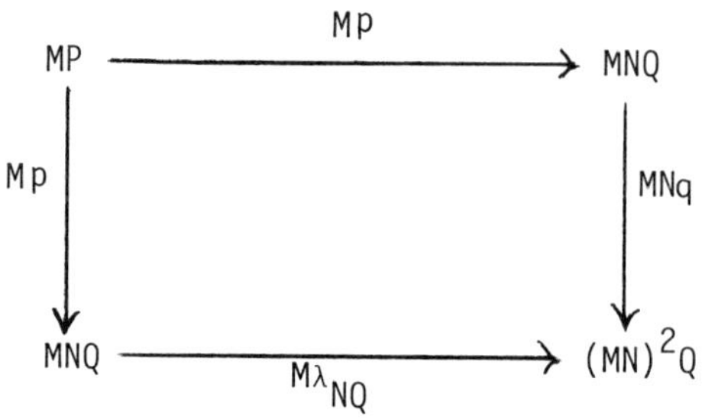

(4)

is an equalizer.

Proof. (i) => (ii). One has

$$MN(\rho_Q Mp) \circ M\lambda_P = MN\rho_Q \circ M(NMp \circ \lambda_P) = MN\rho_Q \circ M\lambda_{NQ}p = MN\rho_Q \circ M\lambda_{NQ} \circ Mp = Mp.$$

Thus the square in (ii) commutes with x the stable isomorphism $\rho_Q Mp$.

(ii) => (iii). Since $M\lambda_P$ is a section of ρ_{MP}, q is a section of ρ_Q. Thus it is enough to show that

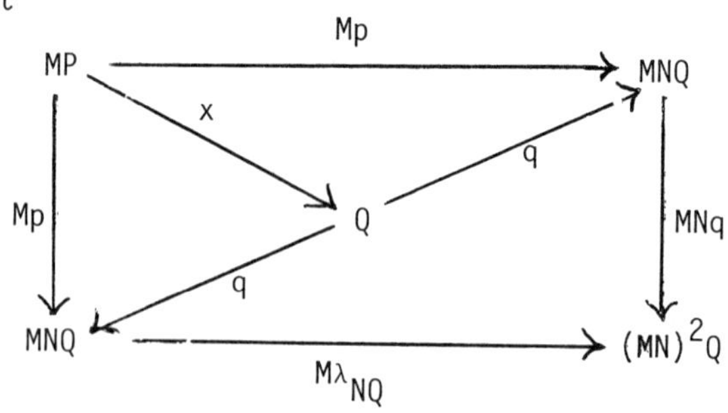

is commutative, where $p = Nx \circ \lambda_P$. But the choice of p entails that $qx = Mp$ so that, in particular, $x = \rho_Q qx = \rho_Q Mp$. But then, by 10.2(i), (3) commutes; and one infers commutativity of the requisite diagram.

(iii) => (i). Clear. □

We call a M-coalgebra of the form $(MP, M\lambda_P)$ with P stable *stably cofree*. The foregoing lemma plainly characterizes stable M-coalgebras stably isomorphic to stably cofree M-coalgebras. In fact, it characterizes stable objects which are the underlying objects of such stable M-coalgebras:

COROLLARY 10.4. A stable object Q of $\mathcal{Q}$ is the underlying object of a stable M-coalgebra stably isomorphic to a stably cofree M-coalgebra precisely when there is a stable object P of $\mathcal{P}$ and a stable arrow $p: P \to NQ$ of $\mathcal{P}$ such that $\rho_Q Mp$ is an isomorphism. □

COROLLARY 10.5. Provided the comparison $\hat{M}$ is stably replete, the following are equivalent properties of $Q \in \text{stab}\ \mathcal{Q}$.

(i) Q is in the stably replete image of M.

(ii) Q is the underlying object of a stable M-coalgebra.

(iii) There is a $P \in \text{stab}\ \mathcal{P}$ and a $p \in \text{stab}\mathcal{P}(P,NQ)$ such that $\rho_Q Mp$ is an isomorphism.

If, moreover, M is G-cotripleable, Q is in the stably replete image of M exactly when there is a stable section q of ρ_Q and a stable equalizer p of λ_{NQ} and Nq such that $\rho_Q Mp$ is an isomorphism.

Proof. Assuming M to be G-cotripleable, let q be structure map for a stable M-coalgebra with underlying object Q. Then, by PTT for G-categories (see 8.8), there is a stable equalizer p as stated which is preserved by M. □

When (Q,q) is a stable M-coalgebra, an obvious way, referring to Lemma 10.3, of ensuring (4) is an equalizer is for there to be a stably split equalizer p of λ_{NQ} and Nq (i.e., split by stable retractions of λ_{NQ} and p). In this case we say that (Q,q) is *stably split*. The situation is elucidated by

LEMMA 10.6. A stable parallel pair $\cdot \overset{y}{\underset{x}{\rightrightarrows}} \cdot$ in a stably amenable G-category has a stably split equalizer if and only if x has a stable retraction x' rendering commutative the square

$$\begin{array}{ccc} \cdot & \xrightarrow{x'y} & \cdot \\ {\scriptstyle x'y}\downarrow & & \downarrow{\scriptstyle y} \\ \cdot & \xrightarrow[x]{} & \cdot \end{array}$$

Proof. If the square is commutative, since $x'yx'y = x'xx'y = x'y$, $x'y$ is a stable idempotent. Consider the diagram

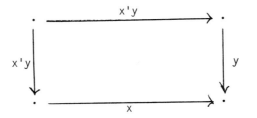

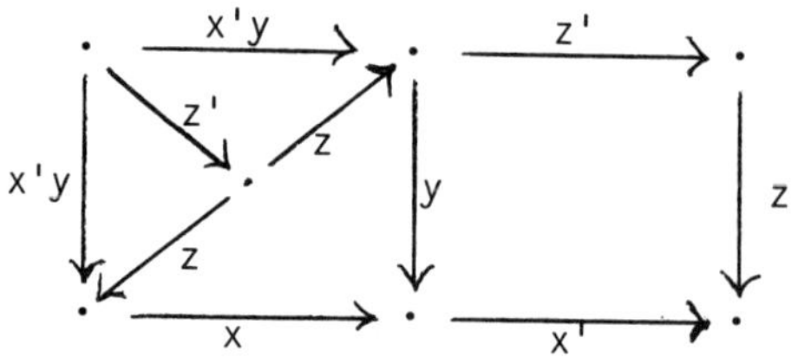

where the right hand square is a stable splitting of x'y. Now, insofar as z' is epi, the diagram is commutative. Thus $\cdot \xrightarrow{z} \cdot \underset{x}{\overset{y}{\rightrightarrows}} \cdot$ is a stably split stable fork. Conversely, if this is a stable fork stably split by x' and z', the displayed diagram is commutative. □

PROPOSITION 10.7. Every stably split stable M-coalgebra is in the stably replete image of $\hat{M}$. If P is stably amenable then

(i) a stable M-coalgebra (Q,q) is stably split if and only if there is a stable retraction r of λ_{NQ} such that $\lambda_{NQ} \circ r \circ Nq = Nq \circ r \circ Nq$, and

(ii) given a stable object Q a necessary and sufficient condition for a stable section q of ρ_Q to be the structure map of a stably split stable M-coalgebra is that, for some stable retraction r of λ_{NQ}, rNq is idempotent and q is the equalizer of M(rNq) and 1.

Proof. The initial statement is an immediate consequence of 10.3 and (i) is just an instance of 10.6. The necessity of the condition in (ii) then follows by 10.2(ii). For the sufficiency, by the result just cited and 10.3, (Q,q) must be a coalgebra. But since, certainly $q\rho_Q M(rNq) = M(rNq)$, the criterion in (i) for (Q,q) to be stably split is met, by 10.2(i). □

COROLLARY 10.8. Suppose that every stable M-coalgebra is stably split and that P is stably amenable. If Q ≅ MP stably for some stable P, there exists a stable section q of ρ_Q and a stable arrow p: P → NQ having a stable retraction p' such that Mp' ∘ q: Q → MP is an isomorphism. □

Towards the proof of Theorem 10.1:

LEMMA 10.9. Any left adjoint whose unit of adjunction is split monic is of descent type.

Proof. Let M be left adjoint to N with unit λ, and let λ' be a retraction of λ. By naturality of λ', the diagram

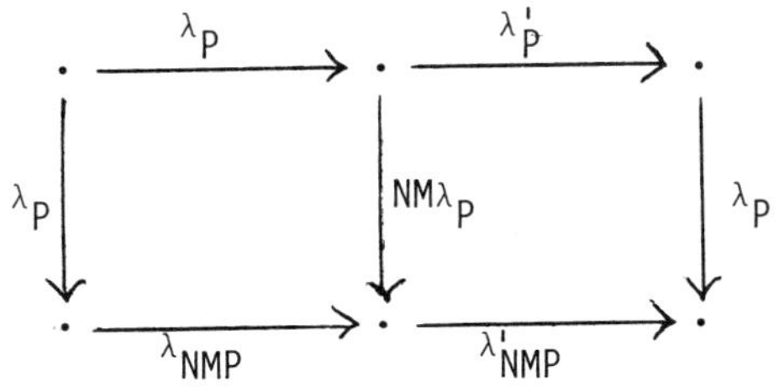

is a split equalizer. In particular, the left square is an equalizer. According to Barr-Wells [2, Corollary 7 and Theorem 9, p. 111], this suffices to prove the result. □

Proof of Theorem 10.1: We may as well assume that G is the trivial group, using 8.7. Thus we can take (1) to be an ordinary adjunction, where idempotents split in P and λ is split monic; and we must prove that M is cotripleable.

Let λ' be a retraction of λ. If, say, (Q,q) is a M-coalgebra then, in the diagram

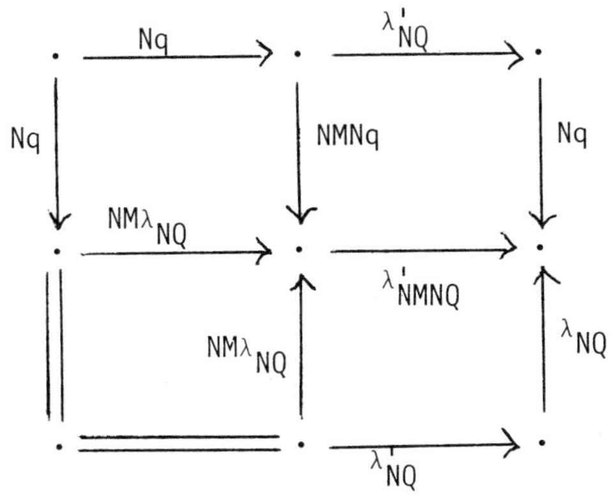

the upper left square commutes by the association law for (Q,q) and the upper and lower right squares commute by naturality of λ'. Hence the diagram is commutative, and we read off the perimeter that $\lambda_{NQ} \circ \lambda'_{NQ} \circ Nq = Nq \circ \lambda'_{NQ} \circ Nq$. Therefore (Q,q) is split, by 10.7(i). But then, M is cotripleable, by 10.7 and 10.9. □

COROLLARY 10.10. If $M: P \to Q$ has a right G-adjoint with unit having a G-natural left inverse then, provided Q is hereditarily stably amenable,

P is hereditarily stably amenable exactly when M is G-cotripleable.

Proof. $\mathcal{Q}$ has H-coequalizers (H < G) of H-stable parallel pairs of the form $\cdot \underset{e}{\overset{1}{\rightrightarrows}} \cdot$, where e is idempotent. So, as follows by 8.1, when M is G-cotripleable P has H-coequalizers of H-stable parallel pairs of the like form. □

COROLLARY 10.11. Let $M \underset{G}{\dashv} N: \mathcal{Q} \to P$ and let λ' be a G-natural retraction of the unit of adjunction and H be a subgroup of G. If P is hereditarily stably amenable, then H-stable coalgebras for M are H-stably split; and H-stable objects Q of $\mathcal{Q}$ are in the H-stably replete image of M precisely when there are H-stable sections q of the counit at Q for which $\lambda'_{NQ}Nq$ is idempotent and the pair $M(\lambda'_{NQ}Nq),1$ has equalizer q.

Proof. See 10.5 and the proofs of 10.1 and 10.7(ii). □

By Corollaries 7.8 and 7.10, Proposition 7.9 and Theorem 10.1, we obtain

THEOREM 10.12. Let C be an amenable G-category. If C has zero arrows, or else C is stably closed and stab C contains an equivalent subcategory of C, stable reflectors for C are cotripleable and their coalgebras are split. In particular, this conclusion is valid when C is an Ab-category (or has a null object). □

A further accounting of this result will be offered at the outset of the next section. Meanwhile:

THEOREM 10.13. Suppose C is an amenable G-category with stable reflections and zero arrows. Denote its stable reflector by R and its respective unit and counit of reflection by μ and ν, and let μ' be the Kronecker retraction of μ (see §7). Assume that stab C has G-fold powers and that, for any $S \in$ stab C, the arrow $RS \to S^G$ induced by the projections from the G-fold power S^G is monic. Then the following are equivalent properties of a stable object D.

(i) D is in the replete image of R.

(ii) There is a monic stable $d:D \to RD$ such that $\mu_D\mu'_Dd = d\mu'_Dd$.

(iii) There is a monic stable $d:D \to RD$ such that $d\nu_DR(\mu'_Dd) = R(\mu'_Dd)$.

Moreover, any monic arrow $d \in \mathrm{stab}\mathcal{C}(D,RD)$ satisfying either of the equations in (ii) or (iii) is the structure map of a split R-coalgebra.

Proof. The implication (i) => (ii) is a consequence of 10.2, 10.7 and 10.5, and (ii) <=> (iii) is immediate by 10.2(i). So assume (ii) and (iii). By (ii), $\mu'_D d$ is an idempotent, because μ'_D is a retraction of μ_D. Thus we may choose an equalizer, say $a: C \to D$, in $\mathcal{C}$ of $\mu'_D d$ and 1; and we have this diagram:

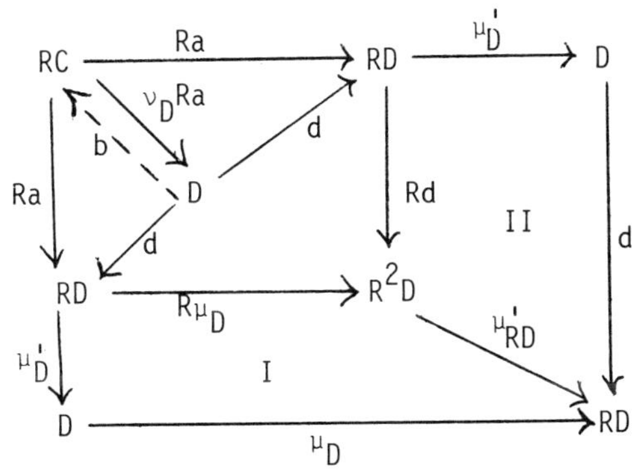

where squares I and II are commutative by naturality of μ'. Also, as follows by 10.6, Ra is an equalizer of $R\mu_D$ and Rd.

Next, whenever $g \in G$

$$^g\mu'_{RD} \circ R\mu_D \circ d = {}^g(\mu'_{RD} \circ R\mu_D \circ d) = {}^g(\mu_D \circ \mu'_D \circ d)$$

$$= {}^g(d \circ \mu'_D \circ d) = {}^g(\mu'_{RD} \circ Rd \circ d) = {}^g\mu'_{RD} \circ Rd \circ d.$$

But, by the definition of μ', ${}^g\mu'_{RD}$ is the projection corresponding to the gth injection for the coproduct $\{{}^g\mu_{RD}: RD \to R^2D\}_{g\in G}$. Hence, by hypothesis, $R\mu_D \circ d = Rd \circ d$. Consequently a pictured stable arrow b exists satisfying $Ra \circ b = d$. This b is monic, by monicity of d. However the equation in (iii) states that $d\nu_D Ra = Ra$; and then, by the properties of equalizers, $b\nu_D Ra = 1$. Thus, $\nu_D Ra$ is an isomorphism. So (D,d) is a split coalgebra for R and D is in the replete image of R, by 10.3. □

As an interesting direct consequence of this result, we refer to Proposition 11.7. For the fruition of the result, see Theorem 13.7.

11. THE CASE OF $\mathcal{D}^G$

First, for trivial G-categories $\mathcal{D}$, we drop the "." in $\mathcal{D}^G$ and simply view $\mathcal{D}^G$ as a G-category via the action of G^{op} on G. The reason, incidentally, for our interest in these $\mathcal{D}^G$ is our interest, exploited in the terminal section after next, in G-categories that are domains of $\mathcal{D}^G$-valued G-cotripleable functors for some $\mathcal{D}$.

The most perspicuous special case of the last-but-one result, Theorem 10.12, of the preceding section is

THEOREM 11.1. Under the assumption that $\mathcal{D}$ is an amenable category with G-indexed coproducts and zero arrows, stable reflectors for $\mathcal{D}^G$ exist and are cotripleable.

The proof is immediate by Corollary 7.5. Notice that if stab $\mathcal{D}^G$ contains an equivalent subcategory of $\mathcal{D}^G$, then all objects of $\mathcal{D}$ are isomorphic. Thus, vis-a-vis Theorem 11.1, the criterion for cotripleability in Theorem 10.12 concerning the stable subcategory is all but vacuous. However, whether or not $\mathcal{D}$ has zero arrows, the same criterion guarantees the existence of a G-category X possessing cotripleable stable reflectors and admitting a G-monadic colimit creating functor $F:X \to \mathcal{D}^G$. This follows, as in the proof of Corollary 10.10, by Theorems 7.4, 7.11 and 8.5. Furthermore, even when there is a G-comonadic such F (see 7.11), stable reflectors on $\mathcal{D}^G$ need not be cotripleable. An example is $\mathcal{D} = \mathrm{Set}^{op}$.

Observe that when $\mathcal{D}$ is the category of pointed sets, cotripleability of stable reflectors on $\mathcal{D}^G$ is an automatic consequence of Theorem 11.1. Yet a critical example occurs when $\mathcal{D}$ = Set. For, despite the fact units of stable reflection are pointwise split monic, coalgebras are not split and Theorem 11.1 cannot apply (by 10.11). Nevertheless stable reflectors are cotripleable, as will appear as an instance of Corollary 11.6 below.

The following notation regarding categories $\mathcal{D}$ with G-indexed coproducts will be retained. We fix, for each object $C = \{C_g\}$ of $\mathcal{D}^G$, a G-indexed coproduct $\iota_g(C):C_g \to LC$. This choice makes L the (object function of the) left adjoint of the diagonal $\Delta:\mathcal{D} \to \mathcal{D}^G$ with unit, μ say, a unit of stable reflection satisfying $(\mu_C)_g = \iota_g(C)$. Thereby the counit ν at $D\varepsilon\mathcal{D}$ is defined by commutativity of

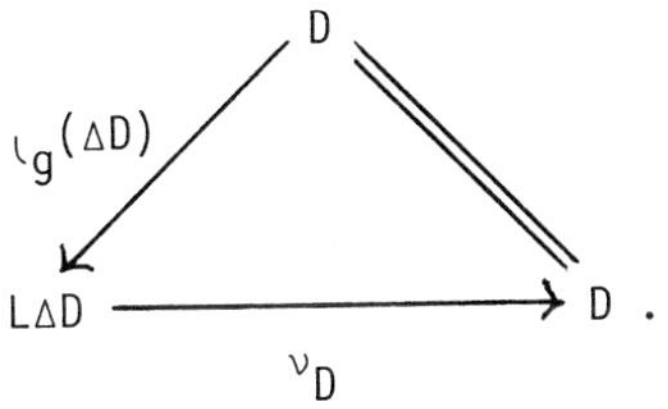

Naturally, cotripleability of L amounts precisely to cotripleability of stable reflectors on $\mathcal{D}^G$.

We call a coalgebra (D,d) for L *semisplit* if there is a G-indexed coproduct of the form $e_g:X_g \to D$ such that e_g is an equalizer of $\iota_g(\Delta D)$ and d for all g. For instance, as the proof of the next result makes apparent, split L-coalgebras are semisplit.

THEOREM 11.2. If for every L-coalgebra (D,d) and every $g\varepsilon G$, the parallel pair $\iota_g(\Delta D), d$ has an equalizer, then the following are equivalent.

(i) L is cotripleable;

(ii) L-coalgebras are semisplit and L reflects isomorphisms;

(iii) μ_C is an equalizer of $\mu_{\Delta LC}$ and $\Delta L\mu_C$ for every $C\varepsilon\mathcal{D}^G$, and every $(D,d) \varepsilon \mathcal{D}_L$ is semisplit.

Proof. Let $(D,d) \varepsilon \mathcal{D}_L$ and consider the commutative diagram

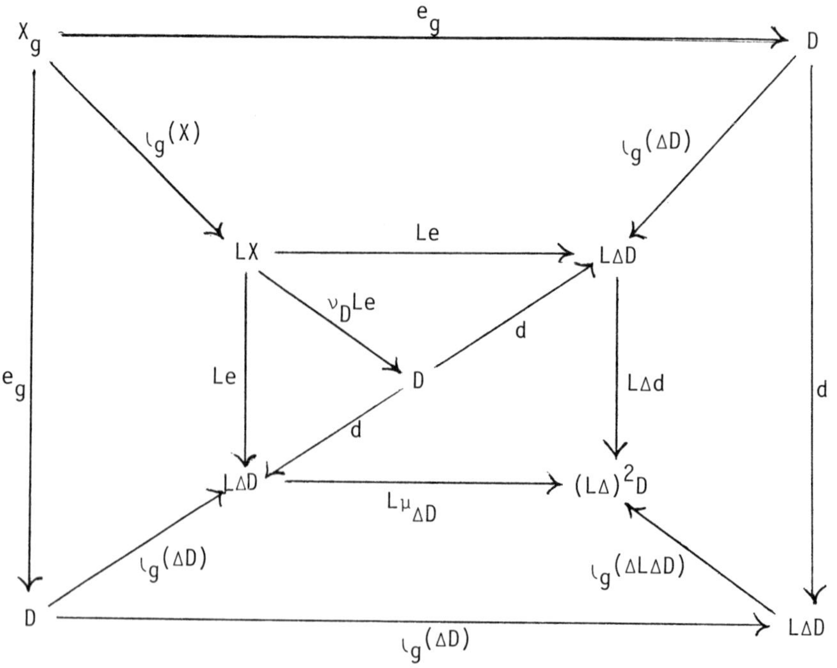

where $X = \{X_g\}$, $e = \{e_g\}$ and the outer square is an equalizer. Since $\nu_D \circ Le \circ \iota_g(X) = \nu_D \circ \iota_g(\Delta D) \circ e_g = e_g$, the e_g form a coproduct if and only if $\nu_D Le$ is an isomorphism; that is, if and only if Le is the equalizer of $L\mu_{\Delta D}$ and $L\Delta d$. Thus by 10.3 and PTT and the fact, previously cited, that μ_C is the equalizer of $\mu_{\Delta LC}$ and $\Delta L\mu_C$ for all C if and only if $\hat{L}$ is fully faithful, it remains to prove the implication (ii) => (iii). For this, assuming (ii), because $(LC, L\mu_C) \in \mathcal{D}_L$ there is a coproduct $\delta_g : Y_g \to LC$ such that δ_g is the equalizer of $\iota_g(\Delta LC)$ and $L\mu_C$. Hence there is a commutative diagram of the form

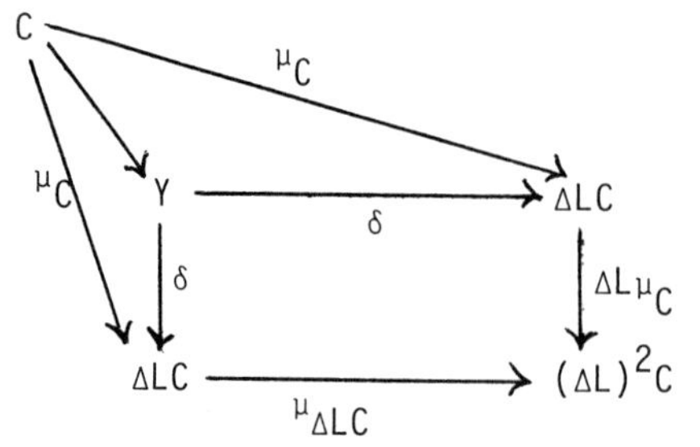

$Y = \{Y_g\}$, $\delta = \{\delta_g\}$

and so, since $L\mu_C$ is an equalizer of $L\mu_{\Delta LC}$ and $L\Delta L\mu_C$, it suffices to show that $L\delta$ is likewise an equalizer of $L\mu_{\Delta LC}$ and $L\Delta L\mu_C$. But this is true by the argument at the start of the proof. □

We remark that if $\mathcal{D}$ has equalizers, it follows by [2, Theorem 9, p.111] that the condition in (iii) concerning equalizers is equivalent to injections for G-indexed coproducts in $\mathcal{D}$ being equalizers (i.e., of parallel pairs). Also, as an immediate consequence of the theorem and Corollary 10.5, we have

COROLLARY 11.3. If L is cotripleable and $\mathcal{D}$ has equalizers, then an object D of $\mathcal{D}$ is in the replete image of L when and only when ν_D has a section d such that any G-indexed family $\{e_g\}$ of equalizers e_g of $\iota_g(\Delta D)$ and d is a family of coproduct injections. □

To make full use of Theorem 11.2 one must be able to tell, given $D\varepsilon\mathcal{D}$, for which sections d of ν_D (if any) the associative law for the pair D,d is valid. This we cannot do in any generality beyond the case of Theorem 11.1 and statement (i) in the next result.

THEOREM 11.4. Suppose $\mathcal{D}$ has equalizers and pullbacks along monics, and that these pullbacks commute with G-indexed coproducts. Then

(i) every L-coalgebra is semisplit and the replete image of L is $\mathcal{D}$, and

(ii) L preserves equalizers whenever injections for the G-indexed coproducts in $\mathcal{D}$ are monic.

Proof. (i). If $(D,d)\ \varepsilon\ \mathcal{D}_L$ then, for each $g\varepsilon G$, there is an equalizer of the form

$$\begin{array}{ccc} X_g & \xrightarrow{\ e_g\ } & D \\ {\scriptstyle e_g}\downarrow & & \downarrow{\scriptstyle d} \\ D & \xrightarrow[\ \iota_g(\Delta D)\ ]{} & L\Delta D \end{array} .$$

But, since $\iota_g(\Delta D)$ and d have the common left inverse ν_D, these equalizers are pullbacks. But then, $e_g:X_g \to D$ is a coproduct. Thus (D,d) is semi-split. Similarly, if D is any object of $\mathcal{D}$, an equalizer of the form

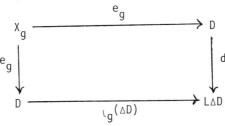

$$X_g \longrightarrow D \underset{\iota_g(\Delta D)}{\overset{\iota_1(\Delta D)}{\rightrightarrows}} L\Delta D$$

is a pullback; so $D \cong LX$.

(ii). Form the commutative diagram (ignore dashed arrows)

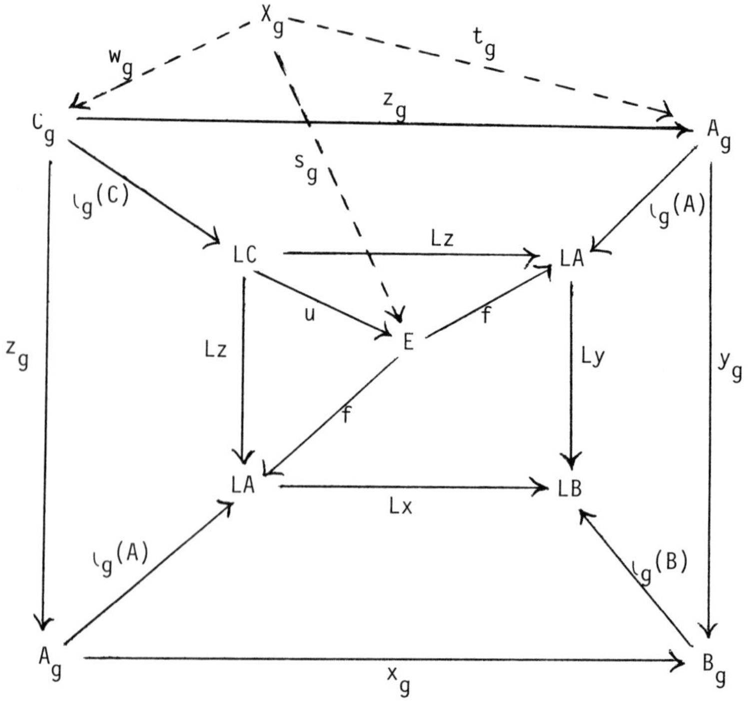

where the outer square is a given equalizer and f is the equalizer of Lx and Ly. Suppose that $fs_g = \iota_g(A)t_g$. Then

$$\iota_g(B) \circ y_g \circ t_g = Ly \circ \iota_g(A) \circ t_g = Ly \circ f \circ s_g$$
$$= Lx \circ f \circ s_g = Lx \circ \iota_g(A) \circ t_g = \iota_g(B) \circ x_g \circ t_g.$$

Therefore $y_g t_g = x_g t_g$. Hence there exists w_g as pictured such that $z_g w_g = t_g$; and then

$$fu\iota_g(C)w_g = Lz\iota_g(C)w_g = \iota_g(A)z_g w_g = \iota_g(A)t_g = fs_g.$$

Thus since f and z_g, being equalizers, are monic, the diagram

$$\begin{array}{ccc}
C_g & \xrightarrow{z_g} & A_g \\
{\scriptstyle \iota_g(C)}\downarrow & & \\
LC & & \downarrow{\scriptstyle \iota_g(A)} \\
{\scriptstyle u}\downarrow & & \\
E & \xrightarrow[f]{} & LA
\end{array}$$

is a pullback. So the $u\iota_g(C)$ form a coproduct and, consequently, u is an isomorphism. □

We point out that for (i) to be valid it is only needed that pullbacks along split monics preserve G-indexed coproducts whenever the pullbacks in question exist. Also, in (ii), this condition ensures that L preserves equalizers of parallel pairs x,y for which Lx,Ly has a split equalizer.

We recall that a coproduct $\xi_i : X_i \to X$ in a category with initial object 0 is disjoint if the ξ_i are monic and the square

$$\begin{array}{ccc}
0 & \longrightarrow & X_j \\
\downarrow & & \downarrow{\scriptstyle \xi_j} \\
X_i & \xrightarrow[\xi_i]{} & X
\end{array}$$

is a pullback whenever $i \neq j$.

THEOREM 11.5. Assume that $\mathcal{D}$ has equalizers, coproducts and a terminal object, that pullbacks along monics exist and commute with coproducts, and that the coproduct of a terminal object with itself is disjoint. Then coproducts are disjoint and L reflects equalizers. Moreover, L preserves monics, provided arrows have images, monic epi arrows are isomorphisms and pullbacks along monics preserve cokernel pairs.

Proof. Let 0 be the initial object and 1 the terminal object. Consider coproducts

$$D \xrightarrow{\lambda} D \amalg E \xleftarrow{\rho} E \qquad\qquad D \xrightarrow{l} D \amalg 1 \xleftarrow{r} 1.$$

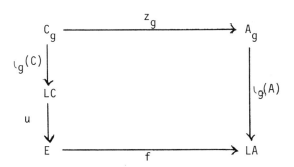

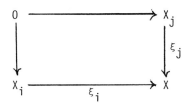

We show the left coproduct is disjoint on the assumption the right one is, this being sufficient for all coproducts to be disjoint. That λ is monic is immediate from the existence of a commutative square

$$\begin{array}{ccc} E & \longrightarrow & 1 \\ {\scriptstyle \rho}\downarrow & & \downarrow{\scriptstyle r} \\ D \amalg E & \xrightarrow[\delta]{} & D \amalg 1 \end{array} \qquad (1)$$

with $\delta\lambda = 1$. We claim this square is a pullback. If it is, inasmuch as pullbacks of monics are monic, ρ is monic. In particular one could form the pullback square on the left in the diagram

$$\begin{array}{ccccc} P & \longrightarrow & E & \longrightarrow & 1 \\ \downarrow & & \downarrow{\scriptstyle \rho} & & \downarrow{\scriptstyle r} \\ D & \xrightarrow[\lambda]{} & D \amalg E & \xrightarrow[\delta]{} & D \amalg 1. \end{array}$$

But the outer rectangle in this diagram would be a pullback too, bearing the implication that $D \amalg E$ is disjoint.

Now, in any event, the squares

$$\begin{array}{ccc} 0 & \longrightarrow & 1 \\ \downarrow & & \downarrow{\scriptstyle r} \\ D & \xrightarrow[\delta\lambda = 1]{} & D \amalg 1 \end{array} \qquad\qquad \begin{array}{ccc} E & \longrightarrow & 1 \\ \Vert & & \downarrow{\scriptstyle r} \\ E & \xrightarrow[\delta\rho]{} & D \amalg 1 \end{array}$$

are pullbacks. Thus since, by hypothesis, pullback along the monic r preserves coproducts, it follows indeed that (1) is a pullback.

To show that L reflects equalizers, consider a commutative diagram of the form

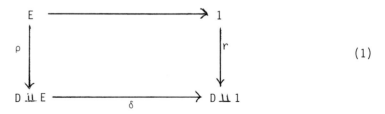

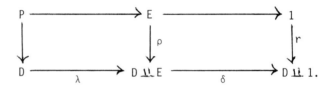

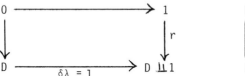

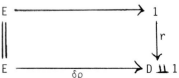

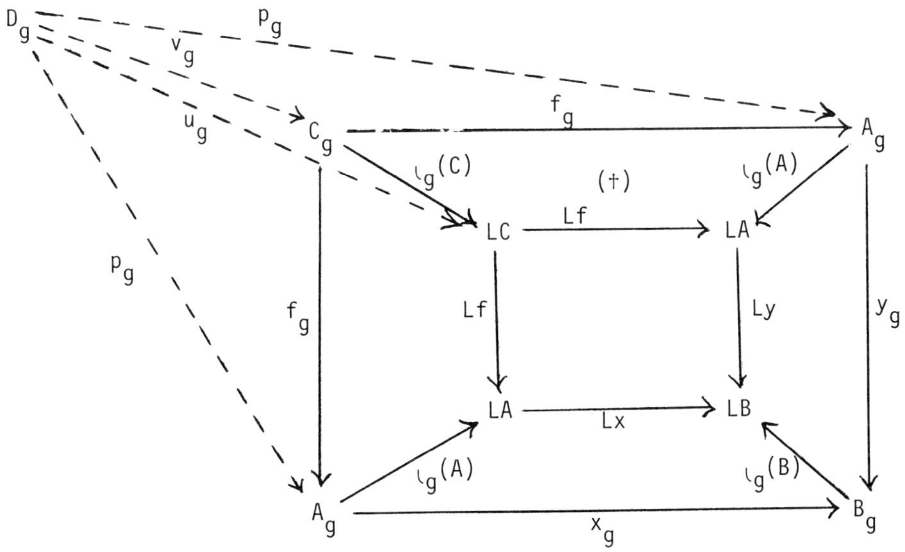

where the innermost square is an equalizer; and suppose, for a moment, that the trapezoid marked (†) is a pullback. It is important to note that, in any case, since disjointness of coproducts has already been demonstrated, f_g is monic (because Lf is). Assuming $x_g p_g = y_g p_g$, one has

$$Lx \circ \iota_g(A) \circ p_g = \iota_g(B) \circ x_g \circ p_g = \iota_g(B) \circ y_g \circ p_g = Ly \circ \iota_g(A) \circ p_g.$$

Hence there exists u_g such that $Lf \circ u_g = \iota_g(A) \circ p_g$. Therefore there is a v_g with $f_g v_g = p_g$. So f_g is the equalizer of x_g and y_g.

As for (†) being a pullback, for fixed g, consider the pullback

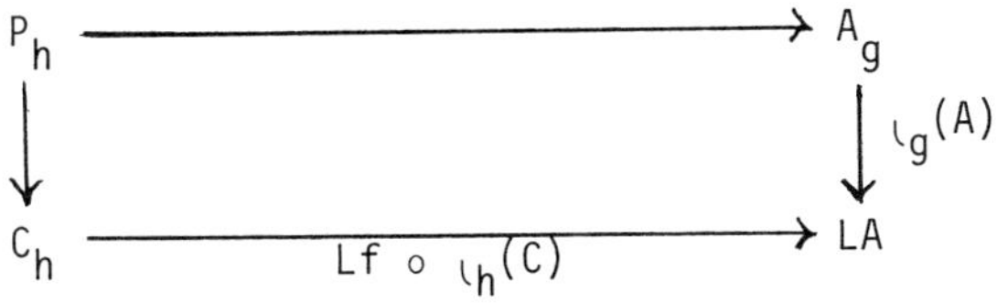

where $h \varepsilon G$. If $h \neq g$, because $Lf \circ \iota_h(C) = \iota_h(A) \circ f_h$ and f_h is monic, $P_h = 0$. So one gets the commutative diagram

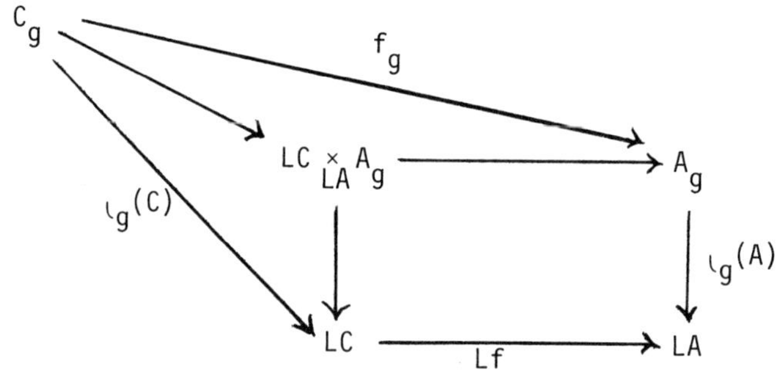

from which it follows easily that (†) is a pullback.

Finally, assume the additional data, and look at the commutative diagram

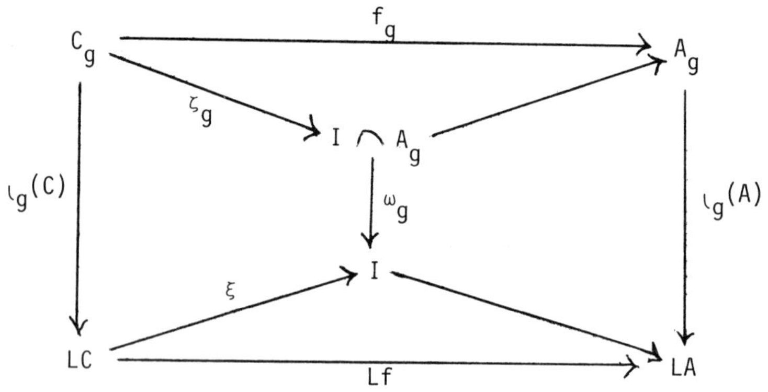

where f_g is monic and the triangle at the bottom is the factorization of Lf through its image. As was just resolved, the outer rectangle in this diagram is a pullback. Thus the square determined by ξ and ω_g is a pullback. But, since $\mathcal{D}$ has equalizers, ξ is epi and, since ω_g is a coproduct injection, ω_g is monic. It follows that ζ_g is epi, hence an isomorphism. Therefore ξ is an isomorphism and Lf is monic. Thus L preserves monics. □

Probably the most significant application of the foregoing is

COROLLARY 11.6. If $\mathcal{E}$ is a complete topos then stable reflectors for $\mathcal{E}^G$ exist and their coalgebras are semisplit. These stable reflectors are replete and cotripleable. Indeed, they preserve monic arrows as well as equalizers of parallel pairs of arrows. □

We refer the reader to Theorem 13.8 for a better result (especially concerning the preservation properties of stable reflectors) and to Theorem 13.10 for a generalization.

We close the section with a determination of the coalgebras in the zero-arrows case:

PROPOSITION 11.7. Suppose $\mathcal{D}$ is amenable with zero arrows and G-fold powers and that, for each $X \in \mathcal{D}$, the evident arrow $G \cdot X \to X^G$ is monic. Then an object D lies in the replete image of L if and only if there is a monic arrow $d: D \to L\Delta D$ such that $\{\iota'_g(\Delta D)d \mid g \in G\}$ is a family of orthogonal idempotents, where $\iota'_g(C)$ denotes the gth projection for the coproduct having injections $\iota_g(C)$.

Proof. Plainly $\mu': \Delta L \to 1$ defined by $(\mu'_C)_g = \iota'_g(C)$ is the Kronecker retraction of μ. Thus, by 10.13, D lies in the replete image of L if and only if, for all g

$$\iota_g(\Delta D)\iota'_g(\Delta D)d = d\iota'_g(\Delta D)d. \tag{2}$$

When these equations hold, the $\iota'_g(\Delta D)d$ are idempotent and, for $h \neq g$

$$\iota'_g(\Delta D)d\iota'_h(\Delta D)d = \iota'_g(\Delta D)\iota_h(\Delta D)\iota'_h(\Delta D)d = 0.$$

Conversely, if the $\iota'_g(\Delta D)d$ are orthogonal idempotents

$$\begin{aligned}\iota'_h(\Delta D)\iota_g(\Delta D)\iota'_g(\Delta D)d &= \iota'_g(\Delta D)d &&\text{if } h = g\\ &= 0 &&\text{if } h \neq g\end{aligned}$$

and

$$\begin{aligned}\iota'_h(\Delta D)d\iota'_g(\Delta D)d &= \iota'_g(\Delta D)d &&\text{if } h = g\\ &= 0 &&\text{if } h \neq g\,.\end{aligned}$$

So the equations (2) hold. □

One should note that pairs D,d satisfying the orthogonal idempotent condition for monic d are, in fact, split L-coalgebras--see Theorem 10.13. Also, for our ultimate result along the lines of Proposition 11.7, see Theorem 13.7 (i).

12. INDUCED STABLE REFLECTIONS AND THEIR SIGNATURES

In applications, reflections of a G-category $\mathcal{A}$ in its stable subcategory occur most often in a more or less disguised form. To elucidate this, we consider a G-functor $H:\mathcal{B} \to \mathcal{A}$ having a left adjoint T and unit of adjunction τ, and a normal subgroup K of G for which H factors through the K-stable subcategory of $\mathcal{A}$. Then $\mathrm{stab}_K\mathcal{A}$ is a G-subcategory of $\mathcal{A}$, so that the corestriction $\overline{H}$ of H to $\mathrm{stab}_K\mathcal{A}$ is a G-functor; and we get a commutative diagram of the form

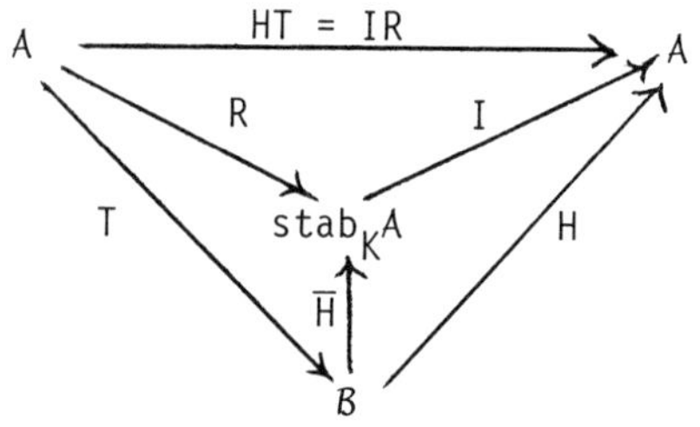

(1)

where I is insertion. Under suitable assumptions on $\overline{H}$--as given by the next result--R will be a reflector of $\mathcal{A}$, seen as K-category, in its stable subcategory; and the unit of reflection will be τ. When this is the case we say that the pair (T,τ) is a *K-reflection over H*. We use the like terminology for reflectors, stable reflections and stable reflectors.

LEMMA 12.1. Referring to the commutative diagram (1): if R is a reflector with given unit and $\overline{H}$ is fully faithful then T is a left adjoint of H having the same unit. Conversely, if T is left adjoint to H with unit τ and $\overline{H}$ has a right adjoint W, then (T,τ) is K-reflective over H if and only if W is full and faithful.

Proof. The first statement follows from the calculation:

$$\mathcal{B}(TA,B) \cong \mathrm{stab}_K\mathcal{A}(\overline{H}TA,\overline{H}B) = \mathrm{stab}_K\mathcal{A}(RA,\overline{H}B)$$
$$\cong \mathcal{A}(A,I\overline{H}B) = \mathcal{A}(A,HB)$$

naturally in A and B. For the second statement, let κ be the counit for the adjunction $T \dashv H$ and η and ε be the respective unit and counit for the adjunction $\bar{H} \dashv W$. Suppose ε is an isomorphism. Then since

$$\mathrm{stab}_K A(R-,-) = \mathrm{stab}_K A(\bar{H}T-,-) \cong \mathcal{B}(T-,W-)$$

$$\cong A(-,HW-) = A(-,I\bar{H}W-) \underset{(I\varepsilon)_*}{\cong} A(-,I-)$$

$R \dashv I$ with unit

$$I\varepsilon\bar{H}T \cdot H\eta T \cdot \tau = (\varepsilon\bar{H} \cdot \bar{H}\eta)T \cdot \tau = 1T \cdot \tau = \tau.$$

On the other hand, suppose $R \dashv I$ with unit τ. Since

$$A(-,I\bar{H}W-) = A(-,HW-) \cong \mathcal{B}(T-,W-)$$

$$\cong \mathrm{stab}_K A(\bar{H}T-,-) = \mathrm{stab}_K A(R-,-) \cong A(-,I-)$$

by Yoneda

$$\varepsilon \cdot H\kappa W \cdot \tau HW = \varepsilon \cdot (H\kappa \cdot \tau H)W = \varepsilon$$

is an isomorphism. □

Now there always is a normal subgroup K such that H factors through $\mathrm{stab}_K A$; namely, the largest subgroup K such that $\mathcal{B}$ is a trivial K-category. This K we call the *stabilizer of* $\mathcal{B}$, and we denote it by $G(\mathcal{B})$. When $G(\mathcal{B})$ is the normal subgroup in question, we say simply that (T,τ) is a *reflection over H*.

Considerations such as in the following result are important in [6].

THEOREM 12.2. Let $E:\mathcal{Y} \to X$ be a G-functor, (S,σ) be a reflection (resp. stable reflection) over E and $F:A \to X$ be a G-functor that creates $G(\mathcal{Y})$-indexed $G(\mathcal{Y})$-coproducts (resp. $G(\mathcal{Y})^{\text{th}}$ copowers).

(i) There is a unique pair (R,μ), where $R:A \to A$ is a functor factoring through $\mathrm{stab}_{G(\mathcal{Y})}A$ and $\mu:1 \to R$ is a natural transformation, such that $FR = ESF$ and $F\mu = \sigma F$. Moreover, viewed as a $\mathrm{stab}_{G(\mathcal{Y})}A$-valued functor, R is a reflector (resp. stable reflector) with unit μ.

(ii) Suppose the inner and outer squares in the diagram

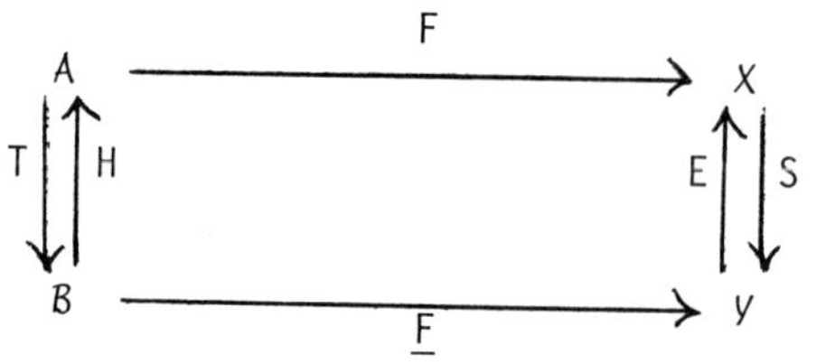

commute, where H and $\underline{F}$ are G-functors and T is a functor, and that

$G(B) \supset G(\mathcal{Y})$. Then FHT = ESF and H factors through $stab_{G(\mathcal{Y})}A$. If, in

addition, there is a natural transformation $\tau{:}1 \to HT$ satisfying $F\tau = \sigma F$ and

the corestriction $\overline{H}$ of H to $stab_{G(\mathcal{Y})}A$ is fully faithful, then (T,τ) is a

$G(\mathcal{Y})$-reflection (resp. $G(\mathcal{Y})$-stable reflection) over H; and $\overline{H}$ is a

G-equivalence provided it has a right G-adjoint.

Proof. (i). By 7.1, for each $A\epsilon A$ the $G(\mathcal{Y})$-indexed family $^g\sigma_{FA}{:}^gFA \to ESFA$

is a $G(\mathcal{Y})$-coproduct. So, for each A, there is a unique pair consisting of

a RA ϵ $stab_{G(\mathcal{Y})}A$ and a family $\{^g\mu_A\}_{g\epsilon G(\mathcal{Y})}$ with $^g\mu_A \epsilon A(^gA,RA)$ such that

$F\mu_A = \sigma_{FA}$; and the $^g\mu_A$ are the injections for a $G(\mathcal{Y})$-coproduct. Plainly

there is a unique way of making R into an endofunctor R on A such that μ_A

is the Ath-component of a natural transformation $\mu{:}1 \to R$. This R, seen as

being $stab_{G(\mathcal{Y})}A$-valued, is a reflector with unit μ, by 7.1.

(ii). The first assertion is elementary and the second assertion is a ready

consequence of (i) and 12.1.

Last, in the case of the parenthetical data, one observes that as in

1.7, a G-functor that creates $G(\mathcal{Y})^{th}$ copowers creates $G(\mathcal{Y})$-stable $G(\mathcal{Y})^{th}$

copowers.   □

Notice that an equivalent (almost) formulation of part (ii) of this

result could be gotten by assuming $G(\mathcal{Y}) = G = G(B)$; that is, $\mathcal{Y}$ and B are

trivial G-categories. Henceforth we take advantage of this simplification.

We wish to study squares

$$\begin{array}{ccc} P & \xrightarrow{\makebox[2cm]{M}} & Q \\ {\scriptstyle H}\Big\uparrow & & \Big\uparrow{\scriptstyle E} \\ \underline{P} & \xrightarrow[\makebox[2cm]{$\underline{M}$}]{} & \underline{Q} \end{array}$$

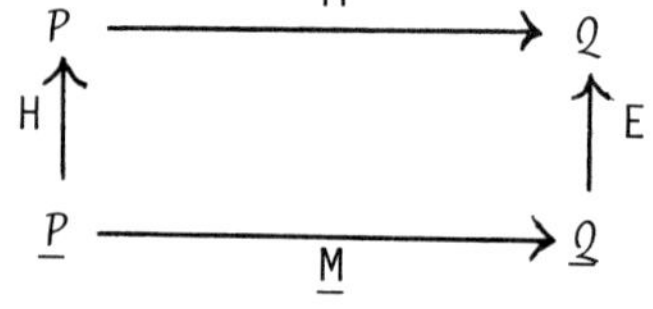

(2)

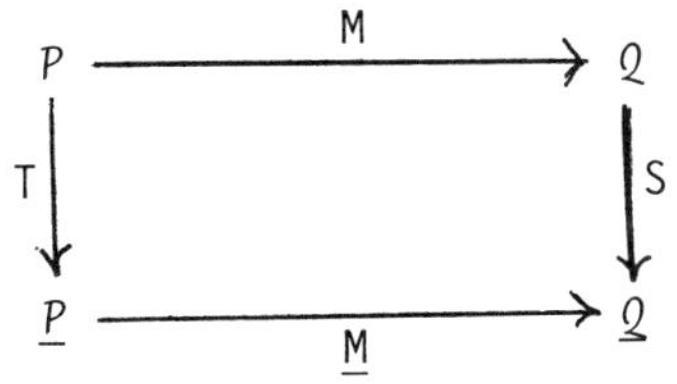

$$\begin{array}{ccc}
P & \xrightarrow{\quad M \quad} & Q \\
{\scriptstyle T}\Big\downarrow & & \Big\downarrow{\scriptstyle S} \\
\underline{P} & \xrightarrow[\underline{M}]{} & \underline{Q}
\end{array} \qquad (3)$$

where M is a given G-cotripleable functor, E and H are given G-functors, S is a given reflector over E, $\underline{M}$ is a G-functor, T is a functor and, typically, the first square commutes and the second one commutes up to a natural isomorphism. We suppose $\underline{P}$ and $\underline{Q}$ to be trivial G-categories. Then (2) commutes and (3) commutes up to isomorphism precisely when, in the diagrams

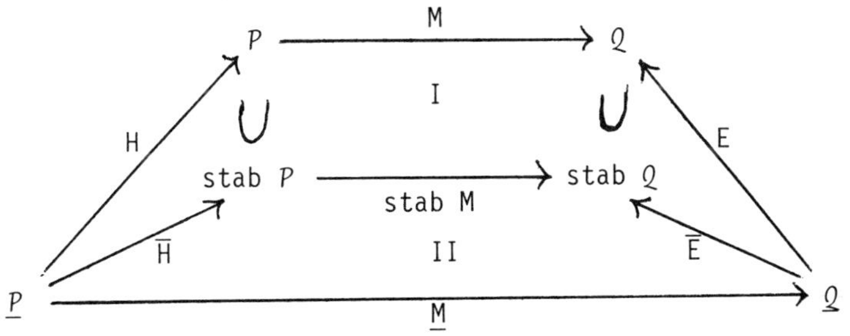

(4)

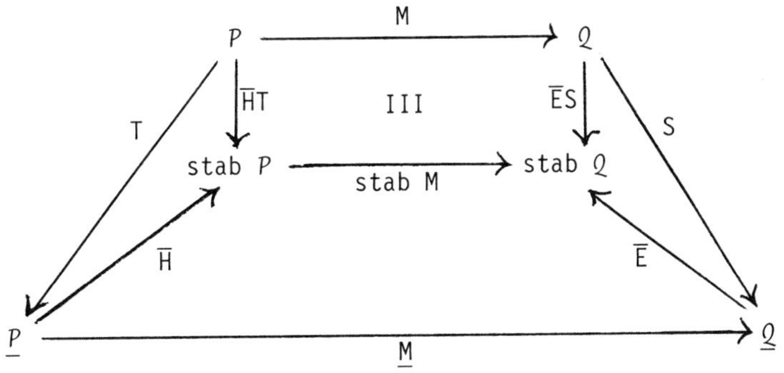

(5)

square II commutes and square III commutes up to isomorphism. In fact, we see more generally that properties of squares (2) and (3) are determined by those of I and III apart from knowledge of the corestrictions $\overline{E}$ and $\overline{H}$. Now the only blanket hypothesis on $\overline{E}$ and $\overline{H}$ ensuring the validity of all that follows is that they are isomorphisms; and so we adopt this as a standing convention--we term E and H *stable embeddings*. (Put differently, a stable embedding of a category $\mathcal{Y}$ in a G-category $\mathcal{X}$ is an embedding of $\mathcal{Y}$ in $\mathcal{X}$ with image stab $\mathcal{X}$.) In some cases better results can be achieved, as was done in Theorem 12.2 (ii), basically by a more detailed analysis of diagrams (4) and (5).

LEMMA 12.3. Let $(M,N;\lambda,\rho):\mathcal{P} \to \mathcal{Q}$ be any G-adjunction. Then there are unique functors $\underline{M}:\underline{\mathcal{P}} \to \underline{\mathcal{Q}}$ and $\underline{N}:\underline{\mathcal{Q}} \to \underline{\mathcal{P}}$ rendering commutative the M-square and N-square in

$$\begin{array}{ccc} \mathcal{P} & \underset{N}{\overset{M}{\rightleftarrows}} & \mathcal{Q} \\ {\scriptstyle H}\uparrow & & \uparrow{\scriptstyle E} \\ \underline{\mathcal{P}} & \underset{\underline{M}}{\overset{\underline{N}}{\leftrightarrows}} & \underline{\mathcal{Q}} \end{array}$$

and a unique adjunction $(\underline{M},\underline{N};\underline{\lambda},\underline{\rho}):\ \underline{\mathcal{P}} \to \underline{\mathcal{Q}}$ (cf. 4.8) such that $H\underline{\lambda} = \lambda H$ and $E\underline{\rho} = \rho E$. Furthermore, if M is G-cotripleable, G-comonadic, a G-equivalence, or injectively G-cotripleable then $\underline{M}$ is respectively cotripleable, comonadic, an equivalence, or injectively cotripleable.

Proof. The existence and uniqueness of $\underline{M}$ and $\underline{N}$ with the requisite properties is a trivial consequence of the assumption that H and E are stable embeddings. Since $\lambda H:H \to NMH$ is a stable natural transformation and $NMH = NE\underline{M} = H\underline{N}\underline{M}$, it follows that $\lambda H = H\underline{\lambda}$ for a unique natural transformation $\underline{\lambda}:1 \to \underline{N}\underline{M}$. Similarly, $\rho E = E\underline{\rho}$ for a unique $\underline{\rho}:\underline{N}\underline{M} \to 1$. Also

$$H(\underline{N}\underline{\rho}\cdot\underline{\lambda}\underline{N}) = H\underline{N}\underline{\rho}\cdot H\underline{\lambda}\underline{N} = NE\underline{\rho}\cdot\lambda H\underline{N}$$

$$= N\rho E\cdot\lambda NE = (N\rho\cdot\lambda N)E = 1.$$

Thus $\underline{N}\underline{\rho}\cdot\underline{\lambda}\underline{N} = 1$ and, similarly, $\underline{\rho}\underline{M}\cdot\underline{M}\underline{\lambda} = 1$.

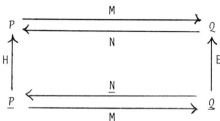

If M is G-cotripleable, $\underline{M}$ is cotripleable by 8.7; and if M is a G-equivalence, $\underline{M}$ is an equivalence by 3.3 (or the above). In particular, as E and H reflect identity arrows, by 8.4 $\underline{M}$ is injectively cotripleable if M is injectively G-cotripleable. Lastly, assume M is G-comonadic, let p be an object of $\underline{P}$, and let $y:\underline{M}p \to q$ be an isomorphism in $\underline{Q}$. Then $Ey:MHp = E\underline{M}p \to Eq$ is an isomorphism in $\mathcal{Q}$, and there exists a unique isomorphism i of the form $Hp \to P$ in $\mathcal{P}$ such that $Mi = Ey$. But $M^g i = {}^gMi = {}^gEy = Ey$ for every $g \in G$; so P and i are stable. Thus there is a unique object p' of $\underline{P}$ and isomorphism $x:p \to p'$ in $\underline{P}$ such that $Hp' = P$ and $Hx = i$. We see that $\underline{M}p' = q$, $\underline{M}x = y$, and we infer that $\underline{M}$ creates isomorphisms. Therefore $\underline{M}$ is comonadic, by 8.4. □

From now on we take M to be G-cotripleable and $\underline{M}$ to be the cotripleable functor making (2) commute.

COROLLARY 12.4. Suppose (3) commutes up to isomorphism and S is cotripleable.
(i) T is cotripleable if and only if $\mathcal{P}$ has and M preserves equalizers of T-contractible equalizer pairs and T has a right adjoint.
(ii) If M preserves equalizers, $\mathcal{P}$ has equalizers of T-contractible equalizer pairs. Also, if $\mathcal{P}$ has equalizers and M preserves equalizers of T-contractible equalizer pairs, then T is cotripleable.

Proof. Let x,y be a T-contractible equalizer pair (i.e., Tx,Ty has a split equalizer).

(i). =>. Let e be an equalizer of x,y. Then Te is an equalizer of Tx,Ty. Thus, since $\underline{M}$ is cotripleable and $\underline{M}$Tx,$\underline{M}$Ty has a split equalizer, $\underline{M}$Te is an equalizer of $\underline{M}$Tx,$\underline{M}$Ty. So SMe is an equalizer of SMx,SMy. But S is cotripleable and SMx,SMy has a split equalizer. Therefore Me is an equalizer of Mx,My.

<=. Let e be an equalizer of x,y. By hypothesis, Me is an equalizer of Mx, My. Consequently SMe is an equalizer of SMx,SMy, since SMx,SMy has a split equalizer and S is cotripleable. Thus $\underline{M}$Te is an equalizer of $\underline{M}$Tx,$\underline{M}$Ty. It follows that Te is an equalizer of Tx,Ty.

Since, plainly, T reflects isomorphisms, one concludes T reflects equalizers for x,y. So T is cotripleable, by PTT.

(ii). If M preserves equalizers, U_M creates equalizers for $\hat{M}x,\hat{M}y$, by 8.2. But, since SMx,SMy has a split equalizer and S is cotripleable, Mx,My has an equalizer. Thus x,y has an equalizer.

Finally, if P has equalizers, T has a right adjoint, by [2, Theorem 3(b), p. 135]. □

In practice, in the situation of this result with T cotripleable (assuming $T \dashv H$ and $S \dashv E$), it is important to note that if $\underline{M}$ has a left adjoint, then M is adjoint tripleable. This follows by the result itself, [2, Theorem 3(a), p. 135] and Corollary 8.9(i).

PROPOSITION 12.5. Assume S is a stable reflector over E. A functor which is a reflector over H is a stable reflector over H; and H is tripleable whenever E is tripleable.

Proof. By 1.7, P has stable G-indexed coproducts. Thus, by 1.3, every reflector over H is a stable reflector. Suppose E is tripleable. Then $\mathcal{Q}$ is stably closed, by 9.7. But then, by 8.5(ii), P is stably closed. So H is tripleable, by 7.1. □

Next we make the standing assumption that S is a reflector over E with unit, say, σ. Also, we denote the counit of the reflection (S,σ) (i.e., the counit for $S \dashv E$) by χ. If (T,τ) is a reflection over H we define its *signature*, with respect to the given reflection (S,σ) over E and G-cotripleable M, to be the natural transformation $\Gamma : SM \to \underline{M}T$ rendering commutative the triangle

$$\begin{array}{ccc} & M & \\ {\scriptstyle \sigma M} \swarrow & & \searrow {\scriptstyle M\tau} \\ ESM & \xrightarrow[E\Gamma]{} & E\underline{M}T = MHT. \end{array} \qquad (6)$$

Notice that, when M is G-comonadic, by Theorem 12.2(i) reflections over H exist and exactly one of them has signature 1.

LEMMA 12.6.(i) If $T : P \to \underline{P}$ is a functor and $\tau : 1 \to HT$ a natural transformation, then (T,τ) is a reflection over H precisely when the natural transformation $\Gamma : SM \to \underline{M}T$ defined by commutativity of the triangle (6) is an isomorphism. In particular, the signature of any reflection over H is a natural isomorphism.

(ii) Given reflections (T,τ) and (T',τ') over H with respective signatures Γ and Γ', $\Gamma' = \underline{M}\eta\cdot\Gamma$ for a unique natural isomorphism $\eta : T \to T'$.

(iii) If κ is the counit for a reflection (T,τ) over H with signature Γ then

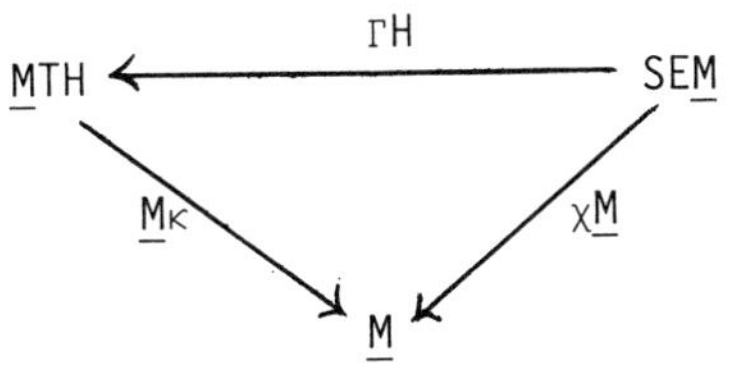

is a commutative triangle.

Proof. (i). By 8.1, M reflects G-indexed G-coproducts and since, by 7.1, $\mathcal{Q}$ has them, M preserves them.

(ii). Let $\eta : T \to T'$ be the natural isomorphism satisfying $H\eta\cdot\tau = \tau'$. Then, as $MH = E\underline{M}$ (see 12.3)

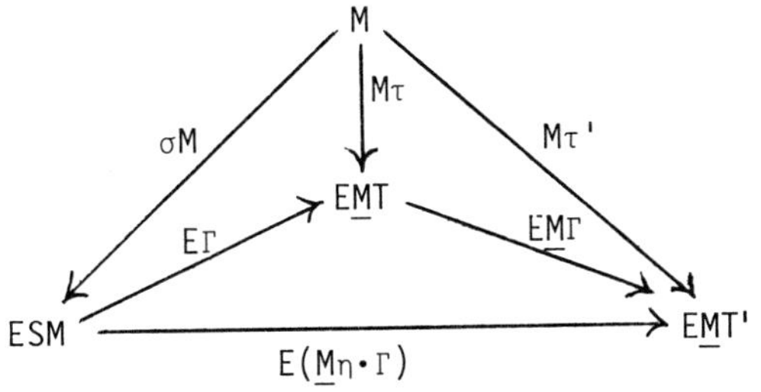

commutes. So $\Gamma' = \underline{M}\eta\cdot\Gamma$. But Γ is an isomorphism, by (i) and, by 8.1 $\underline{M}$, being cotripleable, must be faithful. The uniqueness assertion is now immediate.

(iii). The natural transformation $\chi\underline{M}$ is defined by the equation $E\chi\underline{M}\cdot\sigma E\underline{M} = 1$. However

$$E(\underline{M}\kappa\cdot\Gamma H)\cdot\sigma E\underline{M} = E\underline{M}\kappa\cdot E\Gamma H\cdot\sigma E\underline{M} = MH\kappa\cdot E\Gamma H\cdot\sigma MH$$

$$= MH\kappa\cdot(E\Gamma\cdot\sigma M)H = MH\kappa\cdot M\tau H = M(H\kappa\cdot\tau H) = 1. \quad \square$$

In the light of (ii), we occasionally speak of signatures of reflectors over H, ignoring the units involved. The question of whether or not there are reflectors over H is, incidentally, settled affirmatively in the result after next.

THEOREM 12.7. There is a unique stable embedding $E_M : \underline{\mathcal{Q}}_M \to \mathcal{Q}_M$ such that the diagram

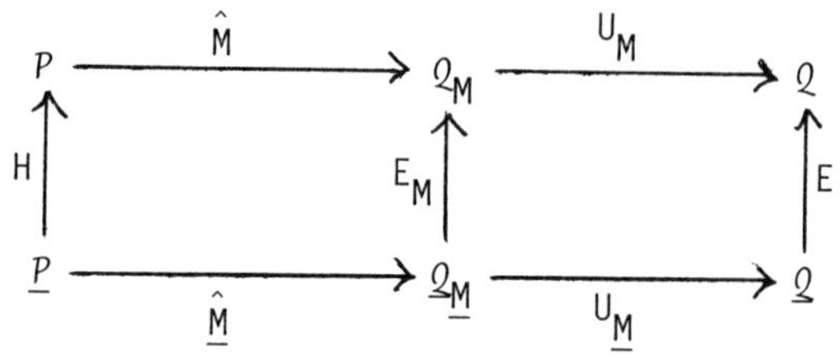

commutes (i.e., the top and bottom rows are the respective standard factorizations of M and $\underline{M}$--see §8).

Proof. Consider the diagram

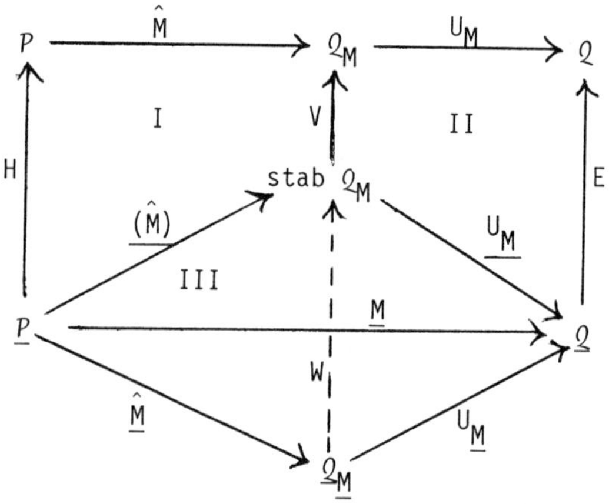

where V is inclusion. Squares I and II and triangle III commute, by 12.3 (ignore the dashed arrow). Thereto, by the same result, $\underline{(\hat{M})}$ is an equivalence and $\underline{U_M}$ is comonadic. Thus, by 8.15, there exists an isomorphism W--the dashed arrow--making, in effect, the whole diagram commutative. In particular, $E_M = VW$ is a stable embedding. That this E_M is as specified is immediate by 8.15(i), since $\underline{\hat{M}}$ is an equivalence and U_M is comonadic. ◻

We denote by (S_M, σ_M) the reflection over E_M such that $U_M E_M S_M = ESU_M$ and $U_M \sigma_M = \sigma U_M$ given by Theorem 12.2(i).

THEOREM 12.8. If (T, τ) is a reflection over H with signature Γ, then S_M is the functor $\mathcal{Q}_M \to \underline{\mathcal{Q}}_M$ such that in the diagram

$$\begin{array}{ccccc} \mathcal{P} & \xrightarrow{\hat{M}} & \mathcal{Q}_M & \xrightarrow{U_M} & \mathcal{Q} \\ {\scriptstyle T}\downarrow & & \downarrow{\scriptstyle S_M} & & \downarrow{\scriptstyle S} \\ \underline{\mathcal{P}} & \xrightarrow[\underline{\hat{M}}]{} & \underline{\mathcal{Q}}_M & \xrightarrow[U_{\underline{M}}]{} & \underline{\mathcal{Q}} \end{array}$$

the right square commutes and the left square commutes up to a unique natural isomorphism $\Gamma_M:S_M\hat{M} \to \underline{\hat{M}}T$ satisfying $U_{\underline{M}}\Gamma_M = \Gamma$. This Γ_M is in fact the signature of (T,τ) with respect to (S_M,σ_M) and $\hat{M}$.

Conversely, choose a G-adjoint equivalence $(\hat{M},K;\omega,\zeta):\mathcal{P} \to \mathcal{Q}_M$, let $(\underline{\hat{M}},\underline{K};\underline{\omega},\underline{\zeta}):\underline{\mathcal{P}} \to \underline{\mathcal{Q}}_M$ be the induced adjoint equivalence of Lemma 12.3, set

$$T = \underline{K}S_M\hat{M}:\mathcal{P} \to \underline{\mathcal{P}} \qquad\qquad \tau = K\sigma_M\hat{M}\cdot\omega:1 \to HT$$

and let $\Gamma:SM \to \underline{M}T$ be defined by commutativity of (6). Then the pair (T,τ) is a reflection over H, Γ is its signature and $\Gamma_M = \underline{\zeta}^{-1}S_M\hat{M}$.

Proof. By 8.14 there is a unique functor $F:\mathcal{Q}_M \to \underline{\mathcal{Q}}_M$ and natural isomorphism $\Gamma_M:F\hat{M} \to \underline{\hat{M}}T$ such that $U_{\underline{M}}F = SU_M$ and $U_{\underline{M}}\Gamma_M = \Gamma$. Put

$$\eta = E_M(F\zeta\cdot\Gamma_M^{-1}K)\cdot\hat{M}\tau K\cdot\zeta^{-1}:1 \to E_MF$$

where we are using the G-adjoint equivalence prescribed in the converse statement. Now since, by naturality of σ

$$\begin{array}{ccc} U_M & \xrightarrow{U_M\zeta^{-1}} & U_M\hat{M}K = MK \\ {\scriptstyle \sigma U_M}\downarrow & & \downarrow{\scriptstyle \sigma MK} \\ ESU_M & \xrightarrow[ESU_M\zeta^{-1}]{} & ESU_M\hat{M}K = ESMK \end{array}$$

commutes, we have

$$U_M\eta = EU_{\underline{M}}(F\zeta\cdot\Gamma_M^{-1}K)\cdot M\tau K\cdot U_M\zeta^{-1} = ESU_M\zeta\cdot E\Gamma^{-1}K\cdot M\tau K\cdot U_M\zeta^{-1}$$

$$= ESU_M\zeta\cdot E\Gamma^{-1}K\cdot(E\Gamma\cdot\sigma M)K\cdot U_M\zeta^{-1} = ESU_M\zeta\cdot\sigma MK\cdot U_M\zeta^{-1} = \sigma U_M.$$

Thus $F = S_M$ and $\eta = \sigma_M$, by 12.2(i). Furthermore since U_M, being comonadic, is faithful, it follows from the definition of Γ that $E_M\Gamma_M\cdot\sigma_M\hat{M} = \hat{M}\tau$; that is, Γ_M is the desired signature.

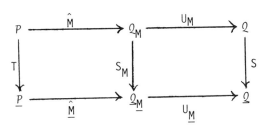

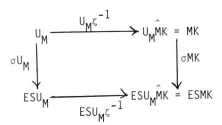

For the converse, observe that

$$\begin{array}{ccccc}
\hat{M} & \xrightarrow{\hat{M}\omega} & \hat{M}K\hat{M} & \xrightarrow{\hat{M}K\sigma_M\hat{M}} & \hat{M}KE_MS_M\hat{M} \\
& \diagdown\!\diagdown & \downarrow{\scriptstyle \zeta\hat{M}} & & \downarrow{\scriptstyle \zeta E_MS_M\hat{M}} \\
& & \hat{M} & \xrightarrow[\sigma_M\hat{M}]{} & E_MS_M\hat{M}
\end{array}$$

is a commutative diagram. Since the vertical composite along the top row is $\hat{M}\tau$ and since, by 12.3, $\zeta^{-1}E_M = E_{\underline{M}}\zeta^{-1}$, applying U_M we get that

$$\begin{array}{ccccc}
& & M & & \\
& \swarrow{\scriptstyle \sigma M} & & \searrow{\scriptstyle M\tau} & \\
ESM & & \xrightarrow[EU_{\underline{M}}\zeta^{-1}S_M\hat{M}]{} & & E\underline{M}T
\end{array}$$

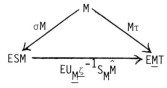

commutes. So (T,τ), Γ and Γ_M are as claimed, by 12.6. □

In practice M can frequently be chosen to be injectively G-cotripleable. This affords a modest simplification vis-a-vis the construction of (T,τ) in the preceding result; for, by Corollary 3.5, to find a G-adjoint equivalence $P \to \mathcal{Q}_M$ it suffices to find a left G-inverse of $\hat{M}$ (and then the unit is 1). Also, one has

PROPOSITION 12.9. If M is injectively G-cotripleable then any two reflections over H having the same signature are identical.

Proof. Let (T,τ) and (T',τ') be reflections over H having a common signature. By the proof of 12.6(ii), the natural isomorphism $\eta: T \to T'$ such that $H\eta \cdot \tau = \tau'$ satisfies $\underline{M}\eta = 1$. But $\underline{M}$ is injectively cotripleable, by 12.3; and thus, by 8.4, $\eta = 1$. □

For the remainder of the section we suppose that (T,τ) is a reflection over H with counit κ and signature Γ. Granted cotripleability of S, when T is cotripleable--see Corollary 12.4--one may try and determine the replete image of T in terms of the coalgebras for S. Thereunder, of course, one must have knowledge of the relationship between coalgebras for T and those for S. In part this is afforded by

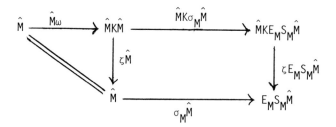

LEMMA 12.10. Given an arrow $\pi: p \to THp$ in $\underline{P}$, (p,π) is a coalgebra for T exactly when $(\underline{M}p, \Gamma^{-1}_{Hp}\,\underline{M}\pi)$ is a coalgebra for S.

Proof. Consider the diagram

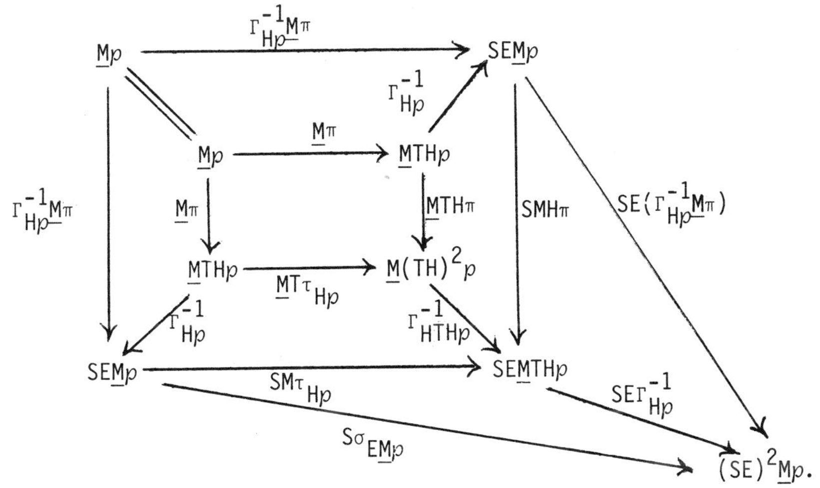

Since $MH = E\underline{M}$ and

$$M\tau_{Hp} = E\Gamma_{Hp} \circ \sigma_{MHp} = E\Gamma_{Hp} \circ \sigma_{E\underline{M}p},$$

by naturality of Γ^{-1}, the outer square commutes if and only if the innermost square commutes. Thus, insofar as $\underline{M}$ is faithful, the outer square commutes if and only if the square

$$\begin{array}{ccc} p & \xrightarrow{\ \pi\ } & THp \\ {\scriptstyle \pi}\downarrow & & \downarrow{\scriptstyle TH\pi} \\ THp & \xrightarrow[T\tau_{Hp}]{} & (TH)^2p \end{array}$$

commutes. But, by 12.6(iii)

$$\kappa_p\pi = 1 \iff M\kappa_p \circ M\pi = 1 \iff \chi_{\underline{M}p} \circ \Gamma^{-1}_{Hp} \circ \underline{M}\pi = 1. \quad \square$$

THEOREM 12.11. Suppose S is cotripleable and that P has and M preserves T-contractible equalizer pairs. Then the following are equivalent properties of an object p of $\underline{P}$.

(i) p is in the replete image of T.

(ii) There is a section π of κ_p such that $(\underline{M}p, \Gamma_{Hp}^{-1}\underline{M}\pi)$ is a S-coalgebra.

(iii) There is an idempotent $e \in \underline{P}(THp, THp)$ such that $\underline{M}e = \Gamma_{Hp} s \chi_{\underline{M}p} \Gamma_{Hp}^{-1}$ for some structure map s for a S-coalgebra with underlying object $\underline{M}p$.

Proof. Since the equivalence of (i) and (ii) is an immediate consequence of the lemma and 10.5, we prove that (ii) is equivalent to (iii).

(ii) => (iii). Take $e = \pi\kappa_p$ and use 12.6(iii).

(iii) => (ii). By 12.6(iii), $\Gamma_{Hp} s \underline{M}\kappa_p = \Gamma_{Hp} s \chi_{\underline{M}p} \Gamma_{Hp}^{-1}$ and $\Gamma_{Hp} s$ is a section of $\underline{M}\kappa_p$. Thus, since $\underline{M}$ reflects coequalizers, it follows there is a section π of κ_p such that $\pi\kappa_p = e$; and then $\underline{M}\pi = \Gamma_{Hp} s$. So $(\underline{M}p, \Gamma_{Hp}^{-1}\underline{M}\pi) \in \underline{\mathcal{Q}}_S$. □

COROLLARY 12.12. With the assumptions of the theorem, suppose in addition S is replete. Then T is replete, provided $\underline{M}$ satisfies the condition: for each $p \in \underline{P}$ and idempotent endomorphism f of $\underline{M}THp$ there is an idempotent endomorphism e of THp such that $\underline{M}e = f$. □

We terminate the section with a result that imparts further information on the interconnections of the coalgebra categories concerned.

THEOREM 12.13. Assume S is cotripleable and P has and M preserves T-contractible equalizer pairs. There is a unique pair W, Ω, where $W: \underline{P}_T \to \underline{\mathcal{Q}}_S$ is a functor and $\Omega: W\hat{T} \to \hat{S}M$ is a natural isomorphism, such that $U_S W = \underline{M}U_T$ and $U_S\Omega = \Gamma$. This W is cotripleable. Moreover, in the diagram

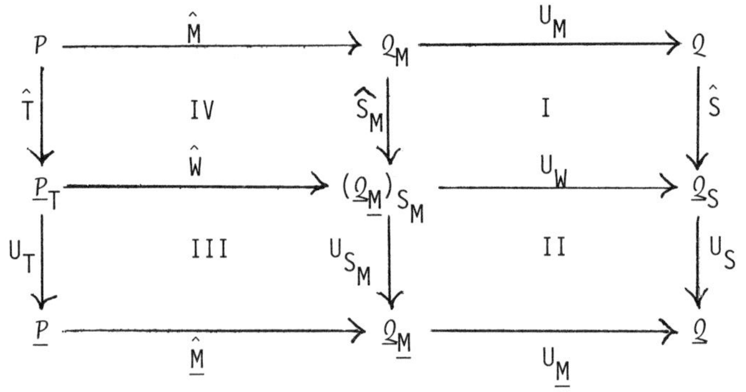

squares I, II and III commute, and square IV commutes up to the natural isomorphism $\Omega_M:\hat{W}\hat{T} \to \widehat{S_M}\hat{M}$ satisfying $U_W\Omega_M = \Omega$. In particular $(\mathcal{Q}_S)_W$ and $(\mathcal{Q}_{\underline{M}})_{S_{\underline{M}}}$ are canonically identifiable categories of coalgebras.

Proof. By 12.4, T is cotripleable. The initial assertion is thus an instance of 8.14.

Consider the diagram

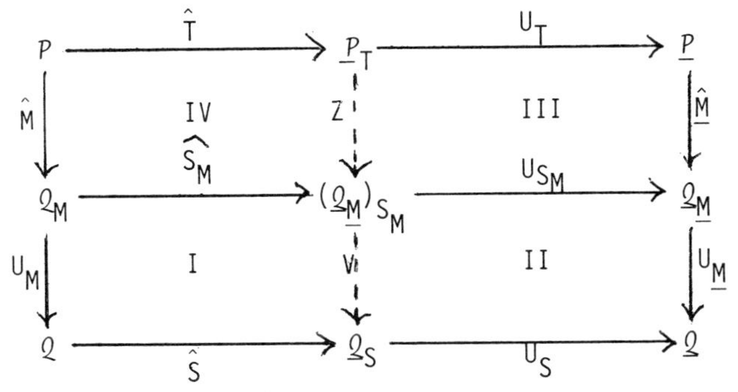

Since $\widehat{S_M}$ is an equivalence and U_S is comonadic, there is a unique V making I and II commute, by 8.14. Since $\hat{T}$ is an equivalence and U_{S_M} is comonadic, by 8.14 there is a unique Z such that III commutes and IV commutes up to the natural isomorphism $\Omega_T:Z\hat{T} \to \widehat{S_M}\hat{M}$ with $U_{S_M}\Omega_T = \Gamma_M$ (see 12.8). But $U_S VZ = \underline{M}U_T$, $U_S V\Omega_T = U_{\underline{M}}U_{S_M}\Omega_T = U_{\underline{M}}\Gamma_M = \Gamma$ and $V\Omega_T:VZ\hat{T} \to \hat{S}M$. Therefore $VZ = W$ and $V\Omega_T = \Omega$.

Now, as IV commutes up to isomorphism, Z is an equivalence. Since I commutes, V is cotripleable, by 8.9(ii). However, since U_{S_M}, $U_{\underline{M}}$ and U_S create isomorphisms, it follows V is comonadic, by 8.4. Hence, by 8.15, W is cotripleable and one may identify Z with $\hat{W}$ and V with U_W. □

13. THE $\mathcal{D}^G$-TARGETED CASE

In this section we study stable reflectors on G-categories admitting $\mathcal{D}^G$-valued G-cotripleable functors. We will consider two cases: when $\mathcal{D}$ is additive and idempotents split, and when $\mathcal{D}$ is a topos. We handle the additive case first, beginning with the definition of an additive G-category.

For a G-category, X say, to be additive, we require not only that it be additive as ordinary category, but that X has a stable null object and the diagonal G-functor $X \to X \times X$ (with the evident G-structure on $X \times X$) has a left G-adjoint. Note that when X is hereditary stably closed, both these additional requirements are vacuous, by Corollary 1.6 and Theorem 3.12. In general, we point out the condition concerning the G-diagonal means there is a coproduct on X of the form say

$$X \xrightarrow{u_{X,Y}} X \amalg Y \xleftarrow{v_{X,Y}} Y$$

where, for all $g \in G$, ${}^g(X \amalg Y) = {}^gX \amalg {}^gY$, ${}^gu_{X,Y} = u_{{}^gX,{}^gY}$ and ${}^gv_{X,Y} = v_{{}^gX,{}^gY}$. In particular we see that each of the natural transformations

$$\Delta_X : X \to X \amalg X \qquad \xi \amalg \xi' : X \amalg X \to Y \amalg Y \qquad \nabla_Y : Y \amalg Y \to Y$$

is G-natural. Thus X is an Ab-category (i.e., ${}^g(\xi + \xi') = {}^g\xi + {}^g\xi'$).

Not surprisingly, the property of being an additive G-category is reflected by many G-cotripleable functors:

PROPOSITION 13.1. If $M: P \to Q$ is G-cotripleable and preserves finite products, and Q is additive, then P is additive.

Proof. We may certainly assume M to be a comonadic G-functor. Then since, by hypothesis, M preserves terminal objects, M creates stable terminal objects, by 8.2 and the proof of 1.7. But M reflects initial objects. Thus P has a stable null object.

Suppose P_1 and P_2 are stable objects of $\mathcal{P}$. Then, as M is a colimit creating G-functor, it follows there is a coproduct

$$P_1 \xrightarrow{\ u_1\ } P_1 \amalg P_2 \xleftarrow{\ u_2\ } P_2$$

in $\mathcal{P}$ such that $P_1 \amalg P_2$, u_1 and u_2 are stable. Put differently, $(P_1 \amalg P_2,\ (u_1,u_2))$ is a stable initial object of the G-category $((P_1,P_2){\downarrow}\Delta)$, where $\Delta:\mathcal{P} \to \mathcal{P} \times \mathcal{P}$ is the diagonal G-functor. But then, by 3.1, this Δ has a left G-adjoint.

We are reduced to showing that $\mathcal{P}$ is an ordinary additive category. For this, let v_i be the ith injection for a finite coproduct in $\mathcal{P}$, and let $v_i^!$ be the corresponding i^{th} projection. So Mv_i is the ith injection for a coproduct in $\mathcal{Q}$ with i^{th} projection $Mv_i^!$. Consequently the $Mv_i^!$ form a product; and hence, by 8.2, the $v_i^!$ form a product. Thus $\mathcal{P}$ has finite biproducts. Therefore $\mathcal{P}$, up to the existence of additive inverses, is an Ab-category. Now, for additive inverses, it is well known to suffice for there to be, given $P \in \mathcal{P}$, $f \in \mathcal{P}(P,P)$ such that $1+f = 0$. Consider the endormorphism of $P \amalg P$ defined by the matrix $a = \begin{bmatrix} 1 & 1 \\ 0 & 1 \end{bmatrix}$. Obviously Ma is invertible. So a is invertible. This entails the existence of an endomorphism f of P with $1+f = 0$. $\square$

Before applying this result to stable reflectors, we need a lemma.

LEMMA 13.2. Let $M:\mathcal{P} \to \mathcal{Q}$ be a functor having a right adjoint N with unit λ having a retraction λ'. If

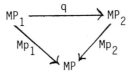

$$\begin{array}{ccc} MP_1 & \xrightarrow{\ q\ } & MP_2 \\ & {\scriptstyle Mp_1}\searrow \quad \swarrow{\scriptstyle Mp_2} & \\ & MP & \end{array}$$

is a commutative triangle in $\mathcal{Q}$, where $p_i \in \mathcal{P}(P_i, P)$, then there is a commutative triangle in $\mathcal{P}$ of the form

$$\begin{array}{ccc} P_1 & \xrightarrow{\ p\ } & P_2 \\ & {\scriptstyle p_1}\searrow \quad \swarrow{\scriptstyle p_2} & \\ & P & \end{array}.$$

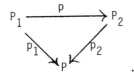

More precisely, the latter triangle commutes with $p = \lambda'_{P_2} \circ Nq \circ \lambda_{P_1}$.

Proof. The diagram

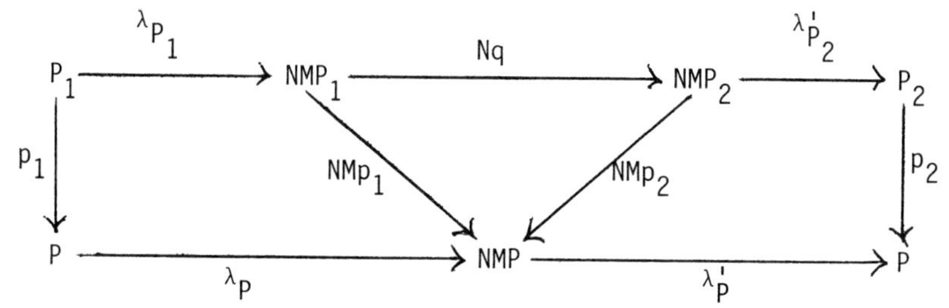

commutes. □

COROLLARY 13.3. The functor M of the preceding lemma reflects limits.

Proof. Let Z be a diagram in *P* and let $\theta : P \to Z$ be a cone such that $M\theta$ is a limiting cone. Suppose $\tilde{\theta} : \tilde{P} \to Z$ is another cone. Then $M\theta \cdot q = M\tilde{\theta}$ for a unique $q : M\tilde{P} \to MP$. By the lemma, $\theta \cdot p = \tilde{\theta}$ for some $p : \tilde{P} \to P$. That there is a unique such p follows from the fact that M is faithful (because λ is pointwise monic). □

We mention that M reflects colimits too.

THEOREM 13.4. Let *C* be a G-category. If *C* is additive, stab *C* is additive. Conversely, if stab *C* is additive and *C* is hereditarily stably closed, amenable and possesses a stable reflector R, the following statements are equivalent.

(i) *C* is additive;

(ii) *C* has an initial object;

(iii) R is cotripleable and preserves terminal objects;

(iv) R is cotripleable, preserves finite products and reflects limits.

Proof. If *C* is additive it follows, by 1.5(iii), that stab *C* has a null object and binary coproducts. Thereto, plainly, stab *C* is an Ab-category under addition of arrows in *C*.

For the converse, suppose stab *C* is additive and assume the given data.

(i) => (ii). Trivial.

(ii) => (iii). Choose a null object *z* of stab *C*. Since insertion of stab *C* in *C* is a right adjoint, *z* is a terminal object of *C*. Thus, by 1.6, *z* is a null object of C. This implies R is cotripleable, by 10.12. In particular, in stab C, Rz is initial and thus terminal. It follows that R preserves terminal objects.

(iii) => (iv). If z is a null object of stab C it is a terminal object of C, so that Rz is an initial object of stab C. Thus C has a null object by cotripleability of R. Hence, by 7.10 and 13.3, R reflects limits. But then, C has finite biproducts, by the argument in 13.1. So R preserves finite products.

(iv) => (i). Since C is hereditarily stably closed, we already know it suffices for C, viewed as ordinary category, to be additive. But this is an immediate consequence of 13.1. □

Next, resurrecting the notation of Section 11, we note that in the terminology of Section 12, L is a stable reflector over the stable embedding $\Delta:\mathcal{D} \to \mathcal{D}^G$ (of course $\mathcal{D}$ has G-indexed coproducts). We fix a G-category A, a G-cotripleable functor $F:A \to \mathcal{D}^G$ and a stable embedding $H:\mathcal{B} \to A$. Thereto we choose a reflection (T,τ) over H having signature Γ--see Theorem 12.8--so that, in particular, the square

$$\begin{array}{ccc} A & \xrightarrow{\quad F \quad} & \mathcal{D}^G \\ {\scriptstyle T}\downarrow & & \downarrow{\scriptstyle L} \\ \mathcal{B} & \xrightarrow[\quad \underline{F} \quad]{} & \mathcal{D} \end{array}$$

commutes up to isomorphism, where $\underline{F}$ is the cotripleable functor making the square

$$\begin{array}{ccc} A & \xrightarrow{\quad F \quad} & \mathcal{D}^G \\ {\scriptstyle H}\uparrow & & \uparrow{\scriptstyle \Delta} \\ \mathcal{B} & \xrightarrow[\quad \underline{F} \quad]{} & \mathcal{D} \end{array}$$

commute (by 12.3).

PROPOSITION 13.5. (i) T is a stable reflector over H, H is tripleable, and A is hereditarily stably closed and has G-indexed coproducts.

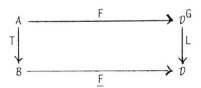

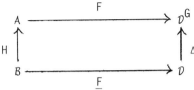

(ii) Any limit of a given type preserved by F is preserved by $\underline{F}$, and any limit or colimit of a given type preserved by L and F is preserved by T.

Proof. (i). Immediate by 7.1, 7.4, 8.5(ii) and 12.5.

(ii). Let Z be a diagram in $\mathcal{B}$ that has a limit. Since H has a left adjoint, HZ has a limit. Thus, on the assumption that F preserves the limit of HZ, FH preserves the limit of Z. So $\Delta\underline{F}$ preserves the limit of Z and therefore, by 1.5(iii), $\underline{F}$ preserves the limit of Z. The assertion concerning limits or colimits preserved by L and F is then a consequence of 8.2 and the fact that cotripleable functors reflect colimits. □

We should mention a point that can be advantageous in practice; namely, if we know beforehand that A is hereditarily stably closed, to show that a functor $A \to \mathcal{D}^G$ is G-cotripleable we need only show, using Theorem 8.5(i), that it is a cotripleable G-functor.

Returning to the additive case:

THEOREM 13.6. Let $\mathcal{D}$ be additive. Then A is additive if and only if F preserves finite products. If $\mathcal{D}$ is amenable, so too is A; and if either A or $\mathcal{B}$ is additive or F preserves finite products, then T is cotripleable, preserves finite products and reflects limits.

Proof. If A is additive it has finite biproducts, whence F preserves finite products. Conversely, if F preserves finite products, A is additive, by 13.1.

Assume $\mathcal{D}$ is amenable. Then A is amenable, by cotripleability of F. Now, by cotripleability of F, A has an initial object and, by 13.4, $\mathcal{B}$ is additive if A is additive. Thus, by 13.5(i) and 13.4, T is as claimed. □

Naturally this result has counterparts for abelian categories, Grothendieck categories, and the like categories $\mathcal{D}$.

As in Section 11, for $C \in \mathcal{D}^G$, we denote the gth injection and g^{th} projection for the coproduct LC of the components C_g of C by $\iota_g(C)$ and $\iota'_g(C)$, respectively. By the preceding two results, Theorems 8.2, 10.13 and 12.11, and Proposition 11.7, we have

THEOREM 13.7. Suppose $\mathcal{D}$ is additive and amenable, and that, for any $D \in \mathcal{D}$, the G-fold power D^G of D exists and the canonical arrow $G\cdot D \to D^G$ is monic.

Assume F preserves finite products and let $B \in \mathcal{B}$.

(i) A necessary and sufficient condition for B to be in the replete image of T is that there be a $b \in \mathcal{B}(B, THB)$ such that $\underline{F}b$ is monic and $\{\iota'_g(\Delta\underline{F}B) \circ \Gamma^{-1}_{HB} \circ \underline{F}b\}_{g\in G}$ is a family of orthogonal idempotents in $\mathcal{D}(\underline{F}B,\underline{F}B)$.

(ii) If F preserves G-fold powers and monic arrows, a necessary and sufficient condition for B to be in the replete image of T is that $\tau_{HB} \circ \tau'_{HB} \circ Hb = Hb \circ \tau'_{HB} \circ Hb$ for some monic $b \in \mathcal{B}(B,THB)$, where τ' is the Kronecker retraction of τ. ◻

We add that the first part of this result becomes more translucent when F is G-comonadic; for then we may take $\Gamma = 1$.

Next we turn to topoi. The following result will be rendered clear by subsequent results in the section.

THEOREM 13.8. Let $\mathcal{D}$ be a topos and assume F preserves equalizers, pullbacks and binary products. Then A and $\mathcal{B}$ are topoi, H is logical, and T is cotripleable and preserves equalizers and pullbacks. Moreover, an object B of $\mathcal{B}$ is in the replete image of T if and only if there is an arrow $b : B \to THB$ of $\mathcal{B}$ such that $\Gamma^{-1}_{HB} \circ \underline{F}b$ is a section of the counit for the stable reflector L over Δ at $\underline{F}B$. ◻

Before going further, if a topos E is a G-category, we demand that its terminal object 1, subobject classifier Ω and true arrow t be stable, and that it be "G-cartesian closed"; that is, the product functor $\times$ is the right G-adjoint of the diagonal $\Delta : E \to E \times E$ and the exponential functor $-^Y$ is the right G(Y)-adjoint of $- \times Y$ for every $Y \in E$. Sometimes we use the term "G-topos". We insist, furthermore, that if a topos valued G-functor Λ on E is logical, Λ preserves G-exponentials in the sense that $\Lambda(X^Y) \cong \Lambda X^{\Lambda Y}$ G(Y)-naturally in X for every Y.

One should know, above, that X^Y can actually be taken to be a G-functor $E^{op} \times E \to E$. As this is merely an instance of a result in [14], we give an informal sketch: for fixed Y and any g, define $X^{gY} = {}^g({}^{g^{-1}}X^Y)$ and $\varepsilon_{gY,X} = {}^g\varepsilon_{Y,g^{-1}X}$, where $\varepsilon_{Y,X} : X^Y \times Y \to X$ is evaluation at Y. Noting $\varepsilon_{gY,X}$ remains

$G({}^gY)$-natural in X, we may assume that ${}^g(X^Y) = {}^gX^{{}^gY}$ and ${}^g\varepsilon_{Y,X} = \varepsilon_{{}^gY,{}^gX}$ for all X,Y and g. Then, defining $X^f : X^{Y_2} \to X^{Y_1}$ for $f : Y_1 \to Y_2$ by commutativity of

$$\begin{array}{ccc} X^{Y_2} \times Y_1 & \xrightarrow{X^f \times 1} & X^{Y_1} \times Y_1 \\ {\scriptstyle 1 \times f}\downarrow & & \downarrow{\scriptstyle \varepsilon_{Y_1,X}} \\ X^{Y_2} \times Y_2 & \xrightarrow[\varepsilon_{Y_2,X}]{} & X \end{array}$$

and applying ${}^g-$, we see we have forced a G(X)-functor structure on each X^- such that $-^-$ is a G-functor.

If $\mathcal{D}$ is a topos, it is elementary that $\mathcal{D}^G$ is a topos and the stable embedding Δ logical. In a less elementary fashion, this is a consequence of the result after next - - it is only indirectly a consequence of the following theorem (cf. 11.6).

THEOREM 13.9. (Lawvere-Tierney). Suppose $N : \mathcal{Q} \to \mathcal{P}$ is a G-functor with left adjoint M, where $\mathcal{Q}$ is a topos and M is of descent type and preserves pullbacks. Let ρ be the counit of the adjunction $M \dashv N$, let Ω be the subobject classifier and t the true arrow of $\mathcal{Q}$, and let χ be the classifying map of MNt.

(i) $N\Omega$ is a subobject classifier and Nt a true arrow for $\mathcal{P}$ if and only if $\rho_\Omega = \chi$.

(ii) Assume $\mathcal{P}$ is hereditarily stably closed and has equalizers, and that M preserves products of pairs of objects.

(a) If $\rho_\Omega = \chi$ then $\mathcal{P}$ is a topos and N is logical.

(b) If M is a left G-adjoint of N, then $\mathcal{P}$ is a topos; and N is logical exactly when $N\chi \circ \lambda_{N\Omega} = 1$, where λ is the unit of the G-adjunction $M \underset{G}{\dashv} N$.

Proof. (i) We have commutative diagrams

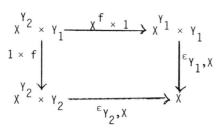

$$\begin{array}{ccc} N1 & = & N1 \\ {\scriptstyle Nt}\downarrow & & \downarrow{\scriptstyle Nt} \\ N\Omega & \xrightarrow{\sigma} & N\Omega \end{array} \qquad \begin{array}{ccccc} MN1 & = & MN1 & \longrightarrow & 1 \\ {\scriptstyle MNt}\downarrow & & \downarrow{\scriptstyle MNt} & & \downarrow{\scriptstyle t} \\ MN\Omega & \xrightarrow{M\sigma} & MN\Omega & \xrightarrow{\rho_\Omega} & \Omega \end{array}$$

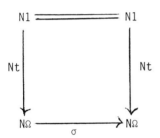

where σ is the endomorphism of $N\Omega$ satisfying $\rho_\Omega \circ M\sigma = \chi$. But the square on the left is a pullback, by 8.2. Thus if Nt = true, $\sigma = 1$ and $\rho_\Omega = \chi$. But, if $\rho_\Omega = \chi$, then $N\chi \circ \lambda_{N\Omega} = 1$, where λ is the unit for $M \dashv N$. Thus, by the proof of a result of Lawvere-Tierney -- see [17, Theorem 2.32, p.54] -- Nt = true.

(ii)-(a). Observe that the diagram

$$\begin{array}{ccc} \mathcal{P} & \xrightarrow{\Delta_{\mathcal{P}}} & \mathcal{P} \times \mathcal{P} \\ {\scriptstyle M}\downarrow & & \downarrow{\scriptstyle M \times M} \\ \mathcal{Q} & \xrightarrow[\Delta_{\mathcal{Q}}]{} & \mathcal{Q} \times \mathcal{Q} \end{array}$$

commutes, Δ_Q has a right adjoint and $M \times M$ is of descent type. Thus, by [2, Theorem 3(b), p.135] and 3.12, $- \times - : \mathcal{P} \times \mathcal{P} \to \mathcal{P}$ is right G-adjoint to $\Delta_{\mathcal{P}}$. In particular, $\mathcal{P}$ is finitely complete, insofar as it has a terminal object. Also, the foregoing argument shows that, for any $P \in \mathcal{P}$, since

$$\begin{array}{ccc} \mathcal{P} & \xrightarrow{- \times P} & \mathcal{P} \\ {\scriptstyle M}\downarrow & & \downarrow{\scriptstyle M} \\ \mathcal{Q} & \xrightarrow[- \times MP]{} & \mathcal{Q} \end{array}$$

commutes apart from natural isomorphism (by hypothesis), $- \times P$ has right $G(P)$-adjoint $-^P$. In addition, for $Q \in \mathcal{Q}$, since $-^Q$ is right $G(Q)$-adjoint to $- \times Q$, $N(-^Q)$ is right $G(Q)$-adjoint to $N(- \times Q) \underset{G(Q)}{\cong} (N-) \times NQ$. So $N(-^Q)$ is

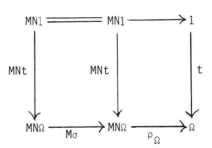

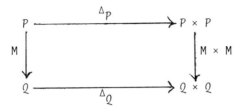

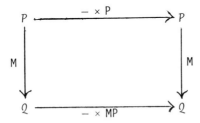

G(Q)-naturally isomorphic to $(N-)^{NQ}$.

(b). Lawvere-Tierney, in the result cited above, construct the subobject classifier in P as the equalizer of the parallel pair

$$N\Omega \xrightarrow{\lambda_{N\Omega}} NMN\Omega \xrightarrow{N\chi} N\Omega$$

$$N\Omega \xrightarrow{\qquad 1 \qquad} N\Omega$$

and the true arrow as the factorization of Nt throught this equalizer (naturality of λ); and their proof requires only that the equalizer exists and that M preserves and reflects pullbacks. Now χ, by its uniqueness, is stable. So the displayed parallel pair is a stable pair, and thus has a stable equalizer. □

Notice that in the proof of this result, the assumption P is hereditarily stably closed was used, outside of producing stable equalizers, only to manufacture G-adjoints "upstairs" (and that without using the G-adjoints downstairs). The extent to which this assumption can be removed is examined in [14]. Relatedly, by the Lawvere-Tierney result (which constructs the exponential as equalizer), (b) remains valid when P is only assumed to have H-stable equalizers for all subgroups H of G, provided M is actually G-cotripleable.

THEOREM 13.10. Let C be a G-category. If C is a topos then

(i) provided monic arrows in stab C remain monic in C and pullbacks of diagrams in stab C of the form

$$\begin{array}{ccc} & & 1 \\ & & \downarrow{\scriptstyle t} \\ \cdot & \longrightarrow & \Omega \end{array}$$

exist and remain pullbacks in C, stab C is a topos and its inclusion functor is logical and

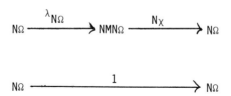

(ii) stable reflectors on C are cotripleable and preserve equalizers and pullbacks. In addition, given a stable reflector, an object of stab C lies in its replete image precisely when the counit of reflection at the object has a stable section.

Conversely, if stab C is a topos and C is G-cartesian closed and has equalizers and cokernel pairs then, provided stable reflectors for C exist and reflect equalizers, C is a topos.

Proof. Assume C is a topos.

(i). By 1.5(iii), 1 is terminal in stab C. Let $m : S_1 \to S_2$ be a monic arrow in stab C. Then m has a classifying map, say χ, in C. But applying ${}^g-$ to the pullback

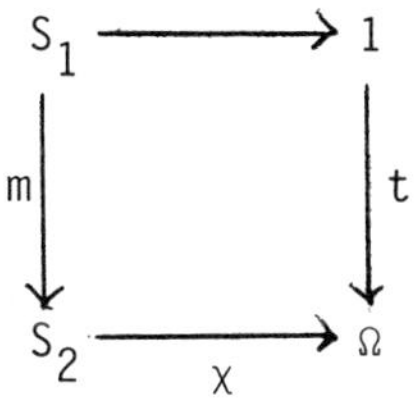

we see χ is stable. Thus this pullback is a pullback in stab C, again by 1.5(iii). It follows that χ is the classifying map of m in stab C. So, in stab C, Ω is the subobject classifier and t = true. It is plain that stab C is cartesian closed with stab $C \subseteq C$ preserving exponentials, by 4.8. In particular, as it is not difficult to show that stab C has equalizers, stab C is finitely complete.

(ii). Let R be a stable reflector on C with respective unit and counit μ and ν. To show that R is cotripleable, we mimic the arguments of Section 11: Since C is a topos, each component μ_C of μ, as an injection for a coproduct, is monic and, being monic, is an equalizer. Thus R is of descent type. Now, letting (D,d) be a R-coalgebra, one has this commutative diagram:

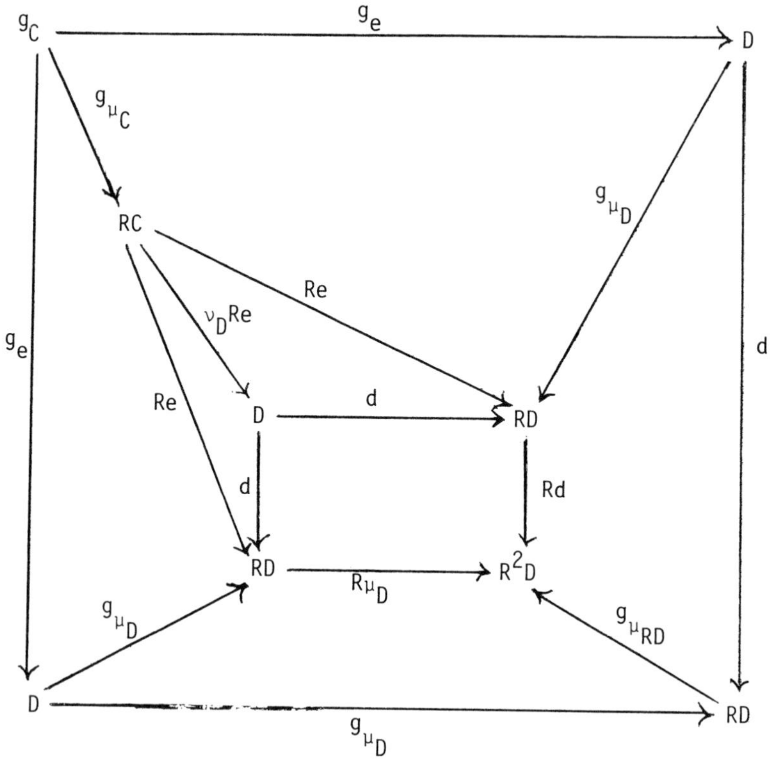

where $e : C \to D$ is the equalizer of μ_D and d in C. Since $\nu_D{}^g\mu_D = {}^g(\nu_D\mu_D) = 1$ and $\nu_D d = 1$, the outer square is a pullback. Thus ${}^ge : {}^gC \to D$ is a coproduct since, in topoi, pullbacks commute with colimits. But $\nu_D \circ Re \circ {}^g\mu_C = \nu_D \circ {}^g\mu_D \circ {}^ge = {}^ge$, so that $\nu_D \circ Re$ is an isomorphism. Thus, by 10.4, (D,d) is isomorphic to a cofree R-coalgebra; establishing cotripleability of R.

It is now clear that any D for which ν_D has a stable section is stably isomorphic to an object of the form RC. Thus since, as in the proof of 11.4 (using 1.5(iii)), R preserves equalizers, it remains to show that R preserves pullbacks. So let

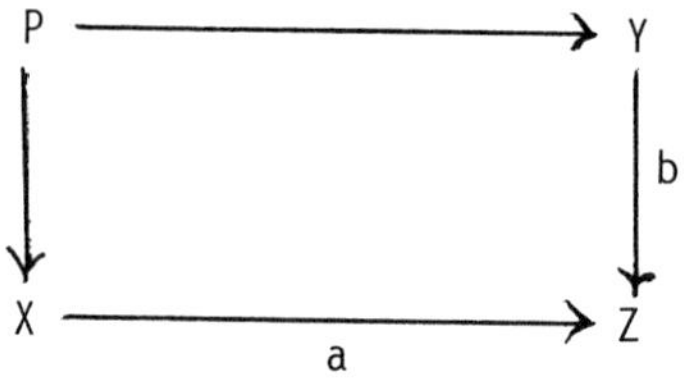

be a pullback in C and, for fixed g and any h, form the pullback

$$\begin{array}{ccc} Q_h & \longrightarrow & {}^hY \\ \downarrow & & \downarrow{\scriptstyle Rb \circ {}^h\mu_Y} \\ {}^gX & \xrightarrow[Ra \circ {}^g\mu_X]{} & RZ \end{array} \quad .$$

Clearly $Q_g \cong {}^gP$. But

$$Ra \circ {}^g\mu_X = {}^g\mu_Z \circ {}^ga, \qquad\qquad Rb \circ {}^h\mu_Y = {}^hb \circ {}^h\mu_Z$$

and, in a topos, coproducts are disjoint and initial objects are strict. It follows that, when $h \neq g$, Q_h is initial. Thus ${}^gX \underset{RZ}{\times} RY \cong {}^gP$. We have

$$RP \cong \coprod {}^gP \cong \coprod ({}^gX \underset{RZ}{\times} RY) \cong \coprod {}^gX \underset{RZ}{\times} RY \cong RX \underset{RZ}{\times} RY \ .$$

We conclude that

$$\begin{array}{ccc} RP & \longrightarrow & RY \\ \downarrow & & \downarrow{\scriptstyle Rb} \\ RX & \xrightarrow[Ra]{} & RZ \end{array}$$

is a pullback in C and hence a pullback in stab C.

Conversely, assume stab C is a topos and C has the properties specified; and let R, μ and ν be as above. Consider a pushout

$$\begin{array}{ccc} \cdot & \xrightarrow{\mu_C} & \cdot \\ {\scriptstyle \mu_C}\downarrow & & \downarrow \\ \cdot & \longrightarrow & \cdot \end{array} \quad .$$

Since R has a right adjoint

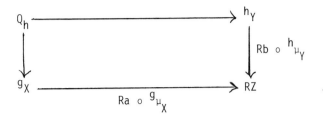

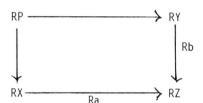

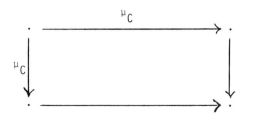

$$\begin{array}{ccc} \cdot & \xrightarrow{R\mu_C} & \cdot \\ {\scriptstyle R\mu_C}\downarrow & & \downarrow \\ \cdot & \longrightarrow & \cdot \end{array}$$

is a pushout and $R\mu_C$ is a coproduct injection. In particular, since stab C is a topos, $R\mu_C$ is the equalizer of its cokernel pair. Thus, as R is assumed to reflect equalizers, μ_C is an equalizer. So R is of descent type.

Since the terminal object of stab C is terminal in C (insertion has a left adjoint), C is a finitely complete cartesian closed category; and such categories are well known to have the property that pullbacks commute with colimits. Thus R is cotripleable, by the proof of (ii). In particular, C has and R reflects initial objects. It follows that, in C, initial objects are strict and coproducts are disjoint. Consequently, by the proof of (ii), R preserves pullbacks.

By 13.9 (i), it now suffices to show that

$$\begin{array}{ccc} R1 & \longrightarrow & 1 \\ {\scriptstyle Rt}\downarrow & & \downarrow{\scriptstyle t} \\ R\Omega & \xrightarrow[\nu_\Omega]{} & \Omega \end{array}$$

is a pullback in C (hence in stab C). We have:

$$R\Omega \underset{\Omega}{\times} 1 \cong G\cdot\Omega \underset{\Omega}{\times} 1 \cong G \cdot (\Omega \underset{\Omega}{\times} 1) \cong G\cdot 1 \cong R1. \qquad \square$$

COROLLARY 13.11. If C is a G-topos with stable reflections, a necessary and sufficient condition for stab C to be a topos and its inclusion functor in C logical is that C be stably closed.

Proof. If C is stably closed, stab C is a finitely complete subcategory of C, by 1.6. On the other hand, suppose stab C is a topos and its insertion logical. But this insertion, being logical, preserves finite colimits. Thus, by

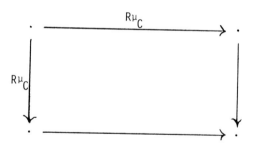

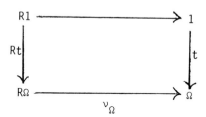

1.5(iii) and PTT, it is tripleable. So $\mathcal{C}$ is stably closed, by 9.7. □

Consider an hereditarily stably closed G-category $\mathcal{C}$ with G-indexed products and coproducts, and let $\mathcal{D}$ be $\mathcal{C}$ seen as ordinary category. Then the functor $U : \mathcal{C} \to \mathcal{D}^G$ given by $(U*)_g = {}^g*$ for every object or arrow $*$ of $\mathcal{C}$ is a G-functor. Moreover, if a parallel pair of arrows x,y in $\mathcal{C}$ is such that Ux, Uy has a split equalizer in $\mathcal{D}^G$, then x,y has a split equalizer in $\mathcal{D}$. It follows, by PTT, that U is G-cotripleable; provided, using Theorem 8.5(i), U has a right adjoint. Now, in many applications, one may show directly U does have a right adjoint, by the adjoint functor theorem. Otherwise, note that the square

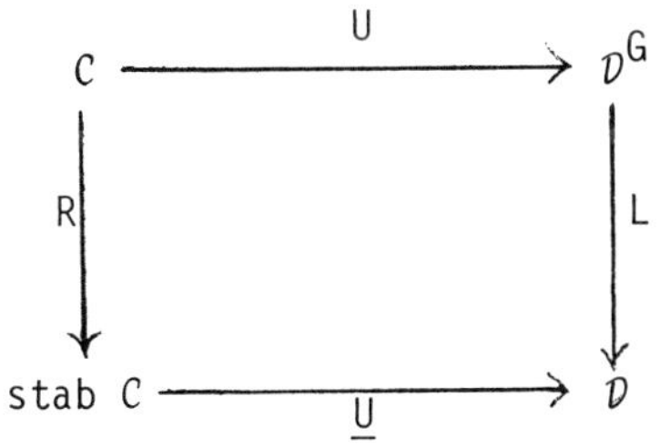

commutes up to isomorphism, where $\underline{U}$ is the evident functor induced by U and R is a stable reflector for $\mathcal{C}$. Thus since, by Theorem 7.1 and Lemma 1.2, it is clear $\underline{U}$ has a right adjoint, U is G-cotripleable whenever $\mathcal{C}$ has equalizers and L is of descent type, by [2, Theorem 3(b), p. 135]. So, by Theorems 11.1 and 13.10, U is G-cotripleable if either $\mathcal{C}$ is additive, amenable and has equalizers or $\mathcal{C}$ is a topos. In fact the assumption $\mathcal{C}$ has equalizers is redundant, according to

LEMMA 13.12. In the diagram

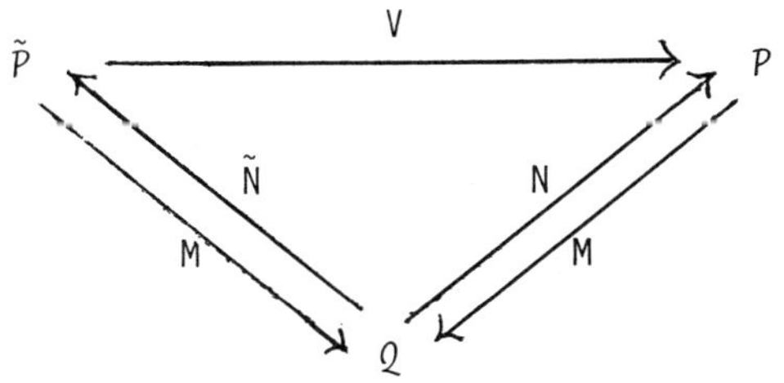

suppose $\tilde{N}$ is right adjoint to $\tilde{M}$, N is right adjoint to M and $MV \cong \tilde{M}$. If $\tilde{\mathcal{P}}$ is amenable and the unit for the adjunction $M \dashv N$ is pointwise split monic, then V has a right adjoint.

Proof. Let $P \in \mathcal{P}$ and set $Q = MP$. By Yoneda, the sequence of natural isomorphisms

$$\tilde{\mathcal{P}}(-,\tilde{N}Q) \cong \mathcal{Q}(\tilde{M}-,Q) \cong \mathcal{Q}(MV-,Q) \cong \mathcal{P}(V-,NQ)$$

provides a terminal object of the form $(\tilde{N}Q,p)$ for the comma category $(V \downarrow NQ)$. Let λ be the unit of adjunction in question and let λ'_P be a retraction of λ_P. Then the diagram

$$\begin{array}{ccc} V\tilde{N}Q & \xrightarrow{\;Ve\;} & V\tilde{N}Q \\ \Big\downarrow{\scriptstyle p} & & \Big\downarrow{\scriptstyle p} \\ NQ & \xrightarrow[\lambda_P\lambda'_P]{} & NQ \end{array}$$

commutes for a unique arrow $e : \tilde{N}Q \to \tilde{N}Q$ in $\tilde{\mathcal{P}}$. Uniqueness of e forces it to be idempotent, because $\lambda_P\lambda'_P$ is idempotent. Thus there is a splitting of e of the form say

$$\tilde{N}Q \xrightarrow{\;\tilde{p}'\;} \tilde{P} \xrightarrow{\;\tilde{p}\;} \tilde{N}Q \quad .$$

But then it follows there is a commutative diagram

$$\begin{array}{ccccc} \tilde{\mathcal{P}}(-,\tilde{N}Q) & \xrightarrow{\;\tilde{p}'_*\;} & \tilde{\mathcal{P}}(-,\tilde{P}) & \xrightarrow{\;\tilde{p}_*\;} & \tilde{\mathcal{P}}(-,\tilde{N}Q) \\ \Big\downarrow & & \Big\downarrow & & \Big\downarrow \\ \mathcal{P}(V-,NQ) & \xrightarrow[(\lambda'_P)_*]{} & \mathcal{P}(V-,P) & \xrightarrow[(\lambda_P)_*]{} & \mathcal{P}(V-,NQ) \end{array}$$

where the vertical arrows are natural isomorphisms. In particular the contravariant functor $\mathcal{P}(V-,P)$ is representable. □

THEOREM 13.13. Let $\mathcal{C}$ be an hereditarily stably closed G-category with G-indexed products and coproducts. If either $\mathcal{C}$ is additive and amenable, or else $\mathcal{C}$ is a topos, there is a G-cotripleable functor $\mathcal{C} \to \mathcal{D}^G$ for some category $\mathcal{D}$ with G-indexed products and coproducts. □

REFERENCES

1. M. Auslander, I. Reiten and S. Smalo, Galois actions on rings and finite Galois coverings, Math. Skand., to appear.

2. M. Barr and C. Wells, "Toposes, Triples and Theories", Grundlehren der Math. Wissenschaften 278, Springer-Verlag, New York, 1984.

3. G.J. Bird, G.M. Kelly, A.J. Power and R.H. Street, Flexible limits for 2-categories, J. Pure and Applied Algebra, to appear.

4. R. Blackwell, G.M. Kelly and A.J. Power, Two-dimensional monad theory, J. Pure and Applied Algebra 59 (1989), 1-41.

5. S. Eilenberg and G.M. Kelly, Closed categories, in "Proc. Conf. on Categorical Algebra", Springer-Verlag, New York (1966), 421-562.

6. Z. Fiedorowicz, H. Hauschild and J.P. May, Equivariant algebraic K-theory, in "Algebraic K-Theory" (R.K. Dennis, Ed.), Lecture Notes in Math. 967, Springer-Verlag, New York (1982), 23-80.

7. Peter Freyd, "Abelian Categories", Harper and Row, New York, 1964.

8. P. Gabriel, The universal cover of a representation-finite algebra, in "Representations of Algebras", (M. Auslander and E. Lluis, Eds.), Lecture Notes in Math. 903, Springer-Verlag, New York (1981), 68-106.

9. R. Gordon, Crossed modules as G-categories, submitted.

10. R. Gordon, Modules over generalized monoids, submitted.

11. R. Gordon, A characterization of G-categories G-equivalent to the G-category of graded modules over a generalized G-graded algebra, J. Algebra, to appear.

12. R. Gordon and E.L. Green, Graded Artin algebras, J. Algebra 76 (1982), 111-137.

13. R. Gordon and E.L. Green, Representation theory of graded Artin algebras, J. Algebra 76 (1982), 138-152.

14. R. Gordon and A.J. Power, G-categories, II, in preparation.

15. E.L. Green, Group-graded algebras and the zero relation problem, in "Representations of Algebras" (M. Auslander and E. Lluis, Eds.), Lecture Notes in Math. 903, Springer-Verlag, New York (1981), 106-115.

16. E.L. Green, Graphs with relations, coverings, and group-graded algebras, Transactions Amer. Math. Soc. 279 (1983), 297-310.

17. P.T. Johnstone, "Topos Theory", Academic Press, London, 1977.

18. G.M. Kelly, "Basic Concepts of Enriched Category Theory", London Math. Soc. Lecture Note Series 64, Cambridge University Press, Cambridge, 1982.

19. G.M. Kelly, Elementary observations on 2-categorical limits, Bulletin Austral. Math. Soc. 39 (1989), 301-317.

20. G.M. Kelly and Ross Street, Review of the elements of 2-categories, in "Category Seminar", Lecture Notes in Math. 420, Springer-Verlag, New York (1974), 75-103.

21. J. Lambek and P.J. Scott, "Introduction to Higher Order Categorical Logic", Cambridge Studies in Advanced Math. 7, Cambridge University Press, Cambridge, 1988.

22. S. Mac Lane, "Categories for the Working Mathematician", Graduate Texts in Math. 5, Springer-Verlag, New York, 1971.

23. Ross Street, Cosmoi of internal categories, Transactions Amer. Math. Soc. 258 (1980), 271-318.

24. J. Taylor, Quotients of groupoids by an action of a group, Math. Proc. Camb. Phil. Soc. 103 (1988), 239-249.

Temple University

Editorial Information

To be published in the *Memoirs*, a paper must be correct, new, nontrivial, and significant. Further, it must be well written and of interest to a substantial number of mathematicians. Piecemeal results, such as an inconclusive step toward an unproved major theorem or a minor variation on a known result, are in general not acceptable for publication. *Transactions* Editors shall solicit and encourage publication of worthy papers. Papers appearing in *Memoirs* are generally longer than those appearing in *Transactions* with which it shares an editorial committee.

As of November 1, 1992, the backlog for this journal was approximately 9 volumes. This estimate is the result of dividing the number of manuscripts for this journal in the Providence office that have not yet gone to the printer on the above date by the average number of monographs per volume over the previous twelve months. (There are 6 volumes per year, each containing about 3 or 4 numbers.)

A Copyright Transfer Agreement is required before a paper will be published in this journal. By submitting a paper to this journal, authors certify that the manuscript has not been submitted to nor is it under consideration for publication by another journal, conference proceedings, or similar publication.

Information for Authors

Memoirs are printed by photo-offset from camera copy fully prepared by the author. This means that the finished book will look exactly like the copy submitted.

The paper must contain a *descriptive title* and an *abstract* that summarizes the article in language suitable for workers in the general field (algebra, analysis, etc.). The *descriptive title* should be short, but informative; useless or vague phrases such as "some remarks about" or "concerning" should be avoided. The *abstract* should be at least one complete sentence, and at most 300 words. Included with the footnotes to the paper, there should be the 1991 *Mathematics Subject Classification* representing the primary and secondary subjects of the article. This may be followed by a list of *key words and phrases* describing the subject matter of the article and taken from it. A list of the numbers may be found in the annual index of *Mathematical Reviews*, published with the December issue starting in 1990, as well as from the electronic service e-MATH [**telnet e-MATH.ams.org** (or **telnet 130.44.1.100**). Login and password are **e-math**]. For journal abbreviations used in bibliographies, see the list of serials in the latest *Mathematical Reviews* annual index. When the manuscript is submitted, authors should supply the editor with electronic addresses if available. These will be printed after the postal address at the end of each article.

Electronically prepared manuscripts. The AMS encourages submission of electronically-prepared manuscripts in $\mathcal{A}\mathcal{M}\mathcal{S}$-TeX or $\mathcal{A}\mathcal{M}\mathcal{S}$-LaTeX. To this end, the Society has prepared "preprint" style files, specifically the amsppt style of $\mathcal{A}\mathcal{M}\mathcal{S}$-TeX and the amsart style of $\mathcal{A}\mathcal{M}\mathcal{S}$-LaTeX, which will simplify the work of authors and of the production staff. Those authors who make use of these style files from the beginning of the writing process will further reduce their own effort.

Guidelines for Preparing Electronic Manuscripts provide additional assistance and are available for use with either AMS-TEX or AMS-LATEX. Authors with FTP access may obtain these *Guidelines* from the Society's Internet node e-MATH.ams.org (130.44.1.100). For those without FTP access they can be obtained free of charge from the e-mail address guide-elec@math.ams.org (Internet) or from the Publications Department, P. O. Box 6248, Providence, RI 02940-6248. When requesting *Guidelines* please specify which version you want.

Electronic manuscripts should be sent to the Providence office only after the paper has been accepted for publication. Please send electronically prepared manuscript files via e-mail to pub-submit@math.ams.org (Internet) or on diskettes to the Publications Department address listed above. When submitting electronic manuscripts please be sure to include a message indicating in which publication the paper has been accepted.

For papers not prepared electronically, model paper may be obtained free of charge from the Editorial Department at the address below.

Two copies of the paper should be sent directly to the appropriate Editor and the author should keep one copy. At that time authors should indicate if the paper has been prepared using AMS-TEX or AMS-LATEX. The *Guide for Authors of Memoirs* gives detailed information on preparing papers for *Memoirs* and may be obtained free of charge from AMS, Editorial Department, P.O. Box 6248, Providence, RI 02940-6248. The *Manual for Authors of Mathematical Papers* should be consulted for symbols and style conventions. The *Manual* may be obtained free of charge from the e-mail address cust-serv@math.ams.org or from the Customer Services Department, at the address above.

Any inquiries concerning a paper that has been accepted for publication should be sent directly to the Editorial Department, American Mathematical Society, P. O. Box 6248, Providence, RI 02940-6248.

Editors

This journal is designed particularly for long research papers (and groups of cognate papers) in pure and applied mathematics. Papers intended for publication in the *Memoirs* should be addressed to one of the following editors:

Ordinary differential equations, partial differential equations, and applied mathematics to JOHN MALLET-PARET, Division of Applied Mathematics, Brown University, Providence, RI 02912-9000

Harmonic analysis, representation theory and Lie theory to AVNER D. ASH, Department of Mathematics, The Ohio State University, 231 West 18th Avenue, Columbus, OH 43210

Abstract analysis to MASAMICHI TAKESAKI, Department of Mathematics, University of California at Los Angeles, Los Angeles, CA 90024

Real and harmonic analysis to DAVID JERISON, Department of Mathematics, M.I.T., Rm 2–180, Cambridge, MA 02139

Algebra and algebraic geometry to JUDITH D. SALLY, Department of Mathematics, Northwestern University, Evanston, IL 60208

Geometric topology, hyperbolic geometry, infinite group theory, and general topology to PETER SHALEN, Department of Mathematics, Statistics, and Computer Science, University of Illinois at Chicago, Chicago, IL 60680

Algebraic topology and differential topology to MARK MAHOWALD, Department of Mathematics, Northwestern University, 2033 Sheridan Road, Evanston, IL 60208-2730.

Global analysis and differential geometry to ROBERT L. BRYANT, Department of Mathematics, Duke University, Durham, NC 27706-7706

Probability and statistics to RICHARD DURRETT, Department of Mathematics, Cornell University, Ithaca, NY 14853-7901

Combinatorics and Lie theory to PHILIP J. HANLON, Department of Mathematics, University of Michigan, Ann Arbor, MI 48109-1003

Logic, set theory, general topology and universal algebra to JAMES E. BAUMGARTNER, Department of Mathematics, Dartmouth College, Hanover, NH 03755

Algebraic number theory, analytic number theory, and automorphic forms to WEN-CHING WINNIE LI, Department of Mathematics, Pennsylvania State University, University Park, PA 16802-6401

Complex analysis and nonlinear partial differential equations to SUN-YUNG A. CHANG, Department of Mathematics, University of California at Los Angeles, Los Angeles, CA 90024

All other communications to the editors should be addressed to the Managing Editor, JAMES E. BAUMGARTNER, Department of Mathematics, Dartmouth College, Hanover, NH 03755.

Recent Titles in This Series

(Continued from the front of this publication)

449 **Michael Slack,** A classification theorem for homotopy commutative H-spaces with finitely generated mod 2 cohomology rings, 1991

448 **Norman Levenberg and Hiroshi Yamaguchi,** The metric induced by the Robin function, 1991

447 **Joseph Zaks,** No nine neighborly tetrahedra exist, 1991

446 **Gary R. Lawlor,** A sufficient criterion for a cone to be area-minimizing, 1991

445 **S. Argyros, M. Lambrou, and W. E. Longstaff,** Atomic Boolean subspace lattices and applications to the theory of bases, 1991

444 **Haruo Tsukada,** String path integral realization of vertex operator algebras, 1991

443 **D. J. Benson and F. R. Cohen,** Mapping class groups of low genus and their cohomology, 1991

442 **Rodolfo H. Torres,** Boundedness results for operators with singular kernels on distribution spaces, 1991

441 **Gary M. Seitz,** Maximal subgroups of exceptional algebraic groups, 1991

440 **Bjorn Jawerth and Mario Milman,** Extrapolation theory with applications, 1991

439 **Brian Parshall and Jian-pan Wang,** Quantum linear groups, 1991

438 **Angelo Felice Lopez,** Noether-Lefschetz theory and the Picard group of projective surfaces, 1991

437 **Dennis A. Hejhal,** Regular b-groups, degenerating Riemann surfaces, and spectral theory, 1990

436 **J. E. Marsden, R. Montgomery, and T. Ratiu,** Reduction, symmetry, and phase mechanics, 1990

435 **Aloys Krieg,** Hecke algebras, 1990

434 **François Treves,** Homotopy formulas in the tangential Cauchy-Riemann complex, 1990

433 **Boris Youssin,** Newton polyhedra without coordinates Newton polyhedra of ideals, 1990

432 **M. W. Liebeck, C. E. Praeger, and J. Saxl,** The maximal factorizations of the finite simple groups and their automorphism groups, 1990

431 **Thomas G. Goodwillie,** A multiple disjunction lemma for smooth concordance embeddings, 1990

430 **G. M. Benkart, D. J. Britten, and F. W. Lemire,** Stability in modules for classical Lie algebras: A constructive approach, 1990

429 **Etsuko Bannai,** Positive definite unimodular lattices with trivial automorphism groups, 1990

428 **Loren N. Argabright and Jesús Gil de Lamadrid,** Almost periodic measures, 1990

427 **Tim D. Cochran,** Derivatives of links: Milnor's concordance invariants and Massey's products, 1990

426 **Jan Mycielski, Pavel Pudlák, and Alan S. Stern,** A lattice of chapters of mathematics: Interpretations between theorems, 1990

425 **Wiesław Pawłucki,** Points de Nash des ensembles sous-analytiques, 1990

424 **Yasuo Watatani,** Index for C^*-subalgebras, 1990

423 **Shek-Tung Wong,** The meromorphic continuation and functional equations of cuspidal Eisenstein series for maximal cuspidal subgroups, 1990

422 **A. Hinkkanen,** The structure of certain quasisymmetric groups, 1990

421 **H. W. Broer, G. B. Huitema, F. Takens, and B. L. J. Braaksma,** Unfoldings and bifurcations of quasi-periodic tori, 1990

420 **Tom Lindstrøm,** Brownian motion on nested fractals, 1990

(See the AMS catalog for earlier titles)